PRACTICAL MENTAL ARITHMETIC

$\Sigma + - \times \div \sqrt{\ } \int$

0 1 2 3 4 5 6 7 8 9

e ln log sin cos tan π

∞

P SCOTT HORNE

Practical Mental Arithmetic

A complete course

HORNE BOOKS

The clarion call of Justice

ISBN-13:
978-1-7382059-4-3 (hardcover)
978-1-7382059-3-6 (paperback)

scott.horne@mail.mcgill.ca
Edmonton, Canada

To
Todd Banovic,
from zero
to infinity

μὴ εἶναι βασιλικὴν ἀτραπὸν ἐπὶ γεωμετρίαν.

There is no royal shortcut to geometry.

—Euclid (answering Ptolemy I)

Contents

Foreword by Ron Doerfler

Classic texts often have simple, direct titles when they encompass a field. The really outstanding ones become known colloquially by their authors' names. The volume you are holding is destined to be one of them; it is a rare and comprehensive survey of fascinating techniques that allow you to achieve seemingly miraculous feats of mental calculation. It is also written with great care and precision. Mr. Horne is a singular authority on the subject as a practiced expert of many years and a devotee of mathematics. To find in that person a talented writer is more than we could rightfully expect; yet there he is, the author of "Horne's book".

The title of this volume exhibits the simple clarity of his prose along with a nod, perhaps, to the primers of old, or to the boards called *hornbooks* that schoolchildren carried as references.* But the subject matter is a modern, approachable analysis of mathematical methods rather than the rote exercises of that time.

* Attributed to English poet William Hornby (1622):

> Even so the Hornbooke is the seede and graine
> Of skill, by which we learning first obtained:
> And though it be accounted small of many,
> And haply bought for twopence or a penny.
> Yet will the teaching somewhat costly be
> Ere they attain unto the full degree
> Of scholarship and art.

Is it "Practical"? Yes, but it is also extraordinary! This book manages to bring into practice a large variety of unusual and often sophisticated methods in a carefully graduated approach. For example, methods to simplify basic multiplications of multi-digit numbers are introduced early on and used consistently throughout. Advanced concepts such as converting denominators into round numbers are presented and repeatedly used within later calculations, whereby they become natural and practical to us. Intermediate steps are patiently shown in detail to avoid any confusion. Only a thoughtfully prepared book can do this effectively.

Is it "Arithmetic"? It's that and much more. We expect to see the basic functions of addition, subtraction, multiplication and division, and perhaps integer powers and roots; all are presented here. These are domains of other past and present lightning calculators, who almost always hew to integer arithmetic. But Mr. Horne also includes mental calculations of decimal powers and roots, along with elementary functions such as logarithms, exponentials and trigonometric functions. These are highly unusual subjects for lightning calculators and exceedingly rare areas for analyses of this kind.

Has it been a while since you've thought about primary-school mathematics? An early chapter serves as an introductory course in arithmetic. Not up to speed on the definitions and use of the elementary functions? Mr. Horne recaps them all. This really is a complete course.

Do you think you know arithmetic? You will be surprised by much of what you see here. The recipes we learned in elementary school for the basic functions (and never thought to revisit) are remnants of ghastly pencil-and-paper exercises. With today's processing power in our hands, we rarely do even those. Let's rise above that: leave the grinding work to electrons, but master the art of visualizing number relationships in concert with learned techniques to produce our *own* calculations, exact or approximate as needs and time permit.

The elements of effective mental calculation are:

1. Methods that are general enough to be effective across wide swaths of problems. This limits the need to learn many rules for very specific cases, a discouraging situation often encountered in books and on online sites. This is where general algebraic formulations of the methods are most effective. Here knowledge of algebra is not required other than knowing that letters in the equations represent non-specific but consistent numbers in a problem. The derivations of the equations can be readily skipped if desired.
2. An absolute minimum number of digits that need to be remembered during the course of the calculation. Memory is fleeting while working on different intermediate calculations. I share the author's preference for announcing a result in left-to-right fashion without writing or calling out intermediate digits, as that is the situation we encounter in daily life. This philosophy means that cross-multiplication, a very common right-to-left approach in competitions, is not ideal here. However, multi-step calculations often do require recollection of intermediate results, and to that end there is an entire chapter of this book dedicated to a standard phonetic system to remember those numbers.
3. A creative and spontaneous knack for recognizing opportunities to exploit the properties of specific numbers to vastly simplify a problem. This is a talent that can be developed (and *will* be developed) over time. Mr. Horne is masterful at teasing us with characteristics of special numbers encountered in the examples he presents, offering us insights into what he would do without burdening us with tables to memorize.

4. Did I mention showmanship? Mr. Horne has a chapter on that!

With practice in the techniques presented here, you will be astonished at what you can do. Some calculations will immediately become simpler once you learn how to approach them in new ways. Others will become easier with practice, and at some point you will realize with growing wonder that your mind is anticipating the next intermediate result as you go through the algorithms. You will develop a flow. Increasingly, you will also recognize familiar numbers and number patterns and instinctively know how to rearrange a difficult problem into a trivial one. In short order you will be surprised when others are impressed by what seems so natural to you. And you will enjoy the thrill of it.

Welcome to the rarefied world of high-precision mental calculation that Mr. Horne offers in this wonderful book. It will last a lifetime.

Ron Doerfler
May 20, 2024

Acknowledgements

Thanks go above all to Ron Doerfler, whose pioneering work on mental and manual calculation inspired this book. He generously read the manuscript just before it went to press and offered a variety of helpful suggestions that have improved the text. His own prior work, especially on such advanced topics as logarithmic and trigonometric functions, is cited several times in these pages, owing to its high quality and great practical utility.

Todd Banovic, Adrienne Camarena, and Sharon Roberts offered their unflagging encouragement and general support, for which I am grateful. Rick Kunst, Canhao Li, and Jack Washam have indulged, or at least tolerated, my mathematical musings for decades. By expressing interest in my work on mental calculation and my writing generally, John Mullen, Yannick Mullen, Grace Parrotta-King, and Rosanna Saccomani helped to inspire me to write this book as a complete course on mental arithmetic.

Stephen Harfield (1981–2023), to my great regret, did not live to see this book. Less than a year ago, when I confided in him that I had in mind to write it, he discouraged me out of a belief that the field of mental arithmetic already abounded in works of great merit and that I would struggle to contribute to the literature. (My view is quite different: most of the literature on mental arithmetic is of poor quality, and no complete textbook existed until I wrote the one in your hands.) He always did value and support my writing, however, and perhaps without presumption I can suppose that he would have seen the finished book in a more favourable light. If not, he might at least have appreciated the justice of my quoting Mao: unless you have investigated the matter, you have no right to speak about it (不做调查没有发言权).

Danielle Howarth of the library at the University of Edinburgh kindly assisted me with research in the archives of Alexander Aitken, which she helps to maintain.

I would be remiss to neglect those of the past few millennia who have developed mental arithmetic, and mathematics generally, to such great heights. Bernard of Chartres observed that we are like dwarfs standing on the shoulders of giants (*nos sumus quasi nani stantes super humeros gigantium*), seeing far not because of any superior ability of our own but because of the towering achievements of those who have gone before us. My own contributions to the grand mathematical edifice do not amount to a cornice here and a mullion there, where others have laid the strong foundation and raised the structure to the skies; this book merely consolidates, streamlines, and teaches what others, some of them nameless, have created, and for those creations they deserve and have my gratitude.

Some of the good people named above have contributed to the book's graces and merits, but its flaws are my responsibility alone.

Introduction

Yes, you can learn to perform in your head difficult calculations that many people would struggle, or fail, to perform on paper: multiplication and division of large numbers, powers and roots, logarithms and exponentials, and trigonometric functions. Far from being innate, skill at mental arithmetic is acquired through diligence and effort. Any intelligent person with the will to try and the motivation to succeed can improve her mental calculating ability, and perhaps attain very great heights indeed.

This book proceeds systematically from rudimentary to complex skills. It begins with a review of conventional arithmetic, for those who may have grown rusty or who, thanks to inferior instruction that has handicapped millions of children and adults, may never have achieved a solid foundation. Even if you do not pursue mental calculation, you may profit from the first chapter, which presents in capsule form the education in arithmetic that few people get in school nowadays, at least in North America.

Many books on mental arithmetic focus on tricks that apply only to special cases. *To multiply by 125, move the decimal point three places to the right and divide by 8.* Yes, that works, for a reason explained in the chapter on fractions (namely, that $125 = 1\,000 \div 8$). But what about multiplying by 119 or 127? No trick will simplify those; they call for *techniques* that apply to broad classes of problems. Here you will learn techniques for multiplying by *any* number, not just a few that happen to underlie flashy tricks. You will also learn tricks for special cases, but in the course of developing general skills; consequently, you will recognise multiplication by 125 on the rare occasion when it arises, and you will be much more likely to notice and use the shortcut than someone who had merely memorised, and promptly forgotten, a list of tricks detached from genuine understanding.

Another gimmick commonly seen in the literature on mental arithmetic is the «calculation» of cube roots. It seems impressive because most people could not extract a cube root even on paper. The gimmick, however, is that the root must be a whole number between 1 and 100. Thanks to a special property of cubes, those roots can be given immediately from a glance at the first few digits and the last digit of the number whose root is sought. In the real world, however, we do not know that the root will come out even; most of the time it will be an endless string of digits, and calculating it takes work. Here, after honing your skills at mental addition, subtraction, multiplication, and division, you will learn how to calculate the cube root of *any* number—in two strikingly different ways—to several decimal places. By contrast, the «skill» of «calculating» a handful of special roots appears nowhere in this book. Those in search of cheap and pretentious gimmicks should look elsewhere.

For the same reason, the examples of calculations presented here were not intended to make the work seem easier than it usually is; most were selected haphazardly, and some were chosen precisely for their difficulty or to illustrate a circumstance that rarely occurs. Several times, when an example turned out to be too easy, I rejected it and chose another. Especially in the early chapters, sometimes the first example given is easy—but the next ones will be more typical.

Like any other skill, mental arithmetic is learnt only through practice. Merely reading this book and admiring the cleverness or efficiency of the techniques presented will not develop your own calculating ability. Please work through the examples yourself, on paper at first if necessary. Develop other problems for practice and check your work with a calculator. If your answer is wrong, repeat the work and find the flaw. *Expect* to make mistakes; welcome them, indeed, as opportunities to correct a weakness. Favour accuracy over speed: an answer obtained however rapidly does no good unless it is correct. Speed will come with facility at performing *accurate* calculations.

Although this book is about calculation itself, in places it connects mental arithmetic to the solution of practical problems, such as conversion of currency and other units of measure, the management of loans and investments, and facts about the physical world. A chapter is devoted to the calendar: using basic skills that you probably already possess, you can determine the day of the week for any date in any year. The chapter on remembering numbers, included to facilitate calculations with more digits than most of us can easily hold in mind, has many practical applications that may improve your life in ways only tangentially related to mathematics.

Arithmetic in these pages is construed broadly enough to encompass powers, roots, logarithms, exponential functions, and trigonometric functions. The first chapter explains all of these in a simple and approachable manner; techniques for computing them appear in some of the final chapters. Tempting though it may be to plunge into that material, you should start at the beginning unless you are overcome with curiosity.

I wish you every success in your study of mental arithmetic and hope that you will acquire useful skills while also gaining a greater understanding and appreciation of mathematics—and having some fun along the way.

P Scott Horne
Edmonton, Alberta, Canada

PRACTICAL MENTAL ARITHMETIC

Chapter 1

A short course in arithmetic

TODAY a firm command of arithmetic shows one's age. Although the jolly boast that «I can't even balance a chequebook» has long conveyed anti-intellectual pride in North America, the schools used to teach rudimentary arithmetic, slowly but passably well. Since the late 1980s, however, they have renounced even that responsibility. Thanks to the inept curricula adopted by «educators», two generations have emerged from high school innocent of the multiplication table and quite unable to perform simple calculations. Allegedly the abandonment of traditional instruction in arithmetic cleared the way for a profound conceptual understanding of mathematics;[1] yet the promise remains unfulfilled, as can be seen when people reach for a calculator in order to multiply by 1.

This chapter offers in capsule form the basic mathematical education that millions of people have never received. You may wish to skim it if you are confident of your command of arithmetic; if you have any doubts, consider studying it carefully. Feel free to skip the more advanced material on powers, roots, loga-

1 Readers of a certain age may remember that the «new math» of the 1960s and early 1970s was promoted on the same basis.

rithms, exponentials, and trigonometric functions if you do not expect to need it.

The decimal number system

Our number system, the *decimal system*, is based on the number 10. We use ten digits, from 0 to 9 (which is $10-1$), to write numbers. The position of a digit indicates its *place value*. The rightmost digit, being in the units' (or ones') place, has a place value of 1; each place has a place value 10 times that of its neighbour to the right. Thus in the number 6 318, the 8 is in the units' place, the 1 is in the tens' place (10×1), the 3 is in the hundreds' place (10×10), and the 6 is in the thousands' place ($10 \times 10 \times 10$). The number thus represents 6 thousands plus 3 hundreds plus 1 ten plus 8 units.

Place value can be represented with an *exponent* that shows the number of factors of the base (here, 10). The number 6 318 is $(6 \times 10^3) + (3 \times 10^2) + (1 \times 10^1) + (8 \times 10^0)$.

This manner of representing numbers can be extended easily to decimal fractions. We write a *decimal point* after the units' place. Since moving to the right divides the place value by 10, the first place after the decimal point is the tenths' place ($\frac{1}{10}$, or 10^{-1}); next is the hundredths' place ($\frac{1}{100}$, or 10^{-2}), and so on.

All of the possible numbers written in this way, with as many digits as necessary on each side of the decimal point, make up the *real numbers*. Numbers whose decimal part is equal to 0 are *integers*.

Decimal representations of numbers are unique, with one exception: a repeating sequence of 9s corresponds to a different representation using a repeating sequence of 0s. The number 1, for example, is both $1.0000\ldots$ and $0.9999\ldots$. The equivalence of these expressions can be seen in different ways. Subtraction yields 0 ($1.0000\ldots - 0.9999\ldots = 0.0000\ldots$), so the numbers are the same. Also, $\frac{1}{9} = 0.1111\ldots$, so $\frac{9}{9} = 9 \times \frac{1}{9} = 0.9999\ldots$; since $\frac{9}{9} = 1$, we see that $1 = 0.9999\ldots$.

It is important to note that the choice of base 10 is an arbitrary convention, one that happens to stem from our having ten fingers. Other bases are possible. Some people advocate base 12 as a better choice than base 10: it uses 12 digits, with values from 0 to 11 (being 12 − 1),[2] and moving to the left multiplies the place value by 12. Modern computers represent numbers internally in the binary system, or base 2, which has only 2 digits (0 and 1). The ancient Maya used base 20 (the vigesimal system) for their sophisticated mathematical work; the people of Bhutan still use it in the dominant Dzongkha language. The Sumerians and the Babylonians used base 60 (the sexagesimal system), which is why we divide an hour into 60 minutes and a minute into 60 seconds, and why we measure angles similarly, with 360 degrees (6 × 60) in a circle.

Fundamental operations

For the real numbers, we treat two operations as fundamental: addition and multiplication. Although addition uses the familiar plus sign (+) from arithmetic, multiplication often replaces the multiplication sign (×) with a centred dot (·), or even with juxtaposition where there is no ambiguity, as in $3a$, xy, $2(5)$, and $4(m + n)$.

Both addition and multiplication are *commutative*: altering the order of the operands does not affect the result. Thus $5 + 4 = 4 + 5$ and $5 \cdot 4 = 4 \cdot 5$. Both operations are also *associative*: three or more operands may be grouped in any way. For instance, the associative property enables us to compute $2+3+4$

2 Commonly the digit representing 10 is written Ɛ and the digit representing 11 is written Ɛ, though other conventions exist.

by first adding $2 + 3$ and then adding 4 to the result (this is written $(2+3)+4$), or by first adding $3+4$ and then adding this result to 2. Both methods yield the same answer, 9. Likewise, $2 \cdot 3 \cdot 4$ can be calculated as $(2 \cdot 3) \cdot 4$ or as $2 \cdot (3 \cdot 4)$; both methods yield 24. Using the associative and commutative properties, we can rearrange addition or multiplication in any convenient way: $a+b+c+d = \big((d+b)+a\big)+c$, and similarly for multiplication.

Multiplication is also *distributive* over addition: $a(b+c) = ab+ac$. For example, to multiply $3 \cdot 12$, we can treat 12 as $10+2$; then, by the distributive property, $3 \cdot 12 = 3(10+2) = 3 \cdot 10 + 3 \cdot 2 = 30 + 6 = 36$.

When added to a number, 0 (the *additive identity*) leaves the number unchanged: $a + 0 = 0 + a = a$. Likewise, a number multiplied by 1 (the *multiplicative identity*) remains the same: $a \cdot 1 = 1 \cdot a = a$.

Zero multiplied by any number is 0.

Negative numbers

Every real number has an *additive inverse*, a unique number that when added to it yields the additive identity, 0. The additive inverse of a is $-a$, called «negative a» or sometimes «minus a». Thus $a + (-a) = (-a) + a = 0$.

As late as the seventeenth century, quite a few Western mathematicians rejected negative numbers as meaningless nonsense. Today, however, even children are familiar with them from temperatures: «six degrees below zero» is −6 °C. A negative balance in an account represents a debt. In golf, a negative result (below par) is better than a positive one (above par). Negative numbers have thus become commonplace in everyday life; they need little explanation.

Subtraction is not a fundamental operation within the real numbers; it is rather the addition of a negative: $a - b = a + (-b)$.

Every real number other than zero is either positive or negative. The negative of a negative is the original number: $-(-a) = a$. Thus subtracting a negative amounts to adding: $7 - (-3) = 7 + \big(-(-3)\big) = 7 + 3 = 10$.

Positive times negative is negative: $2(-3) = 2(-1)(3) = (-1)(2)(3) = -(2 \cdot 3) = -6$. Negative times negative is positive: $(-2)(-3) = (-1)(2)(-1)(3) = -\big(-(2 \cdot 3)\big) = -(-6) = 6$.

The *absolute value* of a number x, written $|x|$, is its value without any negative sign; thus both $|8|$ and $|-8|$ are equal to 8.

Rational numbers

Every real number other than zero has a *multiplicative inverse*, a unique number that when multiplied by it yields the multiplicative identity, 1. The multiplicative inverse of a is $\frac{1}{a}$, also called the *reciprocal* of a. The reciprocal has the same sign as the number itself: $\frac{1}{a}$ is positive when a is positive, negative when a is negative.

Because 0 times any number is 0, there is no multiplicative inverse of 0: no number multiplied by 0 can yield 1. Therefore, 0 has no reciprocal, and the notation $\frac{1}{0}$ is meaningless.

Division ($a \div b$ or $\frac{a}{b}$) is not a fundamental operation within the real numbers; it is defined as multiplication by the reciprocal: $\frac{a}{b} = a \cdot \frac{1}{b}$. Because zero has no reciprocal, division by zero is undefined; it cannot be done.

A *rational number* is a fraction in which one integer is divided by another (other than zero). Every integer, therefore, is rational, because it is equal to itself divided by 1. The reciprocal of a rational number can be formed by exchanging the numerator and the denominator: the reciprocal of $\frac{2}{3}$ is $\frac{3}{2}$; the reciprocal of $\frac{1}{6}$ is $\frac{6}{1}$, or 6.

Any rational number can be written in decimal form, though it may have an infinitely repeating sequence of digits: for example, $\frac{1}{3} = 0.33333\ldots$, and $\frac{7}{22} = 0.3181818\ldots$. The repeating

+	1	2	3	4	5	6	7	8	9
1	2	3	4	5	6	7	8	9	10
2	3	4	5	6	7	8	9	10	11
3	4	5	6	7	8	9	10	11	12
4	5	6	7	8	9	10	11	12	13
5	6	7	8	9	10	11	12	13	14
6	7	8	9	10	11	12	13	14	15
7	8	9	10	11	12	13	14	15	16
8	9	10	11	12	13	14	15	16	17
9	10	11	12	13	14	15	16	17	18

sequence can be marked with a bar over it: $\frac{1}{3} = 0.\overline{3}$, and $\frac{7}{22} = 0.3\overline{18}$. (Even a rational number with a terminating decimal technically has a repeating sequence: $\frac{1}{2} = 0.5 = 0.50000\ldots = 0.5\overline{0}$.)

Numbers whose decimal part does not eventually repeat are *irrational numbers*; they cannot be written exactly with decimals or as fractions, although they can be approximated to as many decimal places as desired. Examples of irrational numbers are $\sqrt{2}$ (the number, approximately 1.414 213 56, that when multiplied by itself gives 2 exactly) and π (the ratio of the circumference of a circle to the diameter, approximately 3.141 592 65).

Addition

Addition requires a thorough knowledge of the addition table, which is reproduced here up to 9+9. Sums involving 0 are omitted, since adding 0 leaves a number unchanged.

To add numbers, begin by aligning them vertically so that digits of the same place value appear in a single column:

$$\begin{array}{r} 4271 \\ 3965 \\ \hline 8236 \end{array}$$

The calculation then proceeds from right to left. In the units' place, 1 + 5 = 6; we write 6 as the units' digit of the sum. In the tens' place, 7 + 6 = 13 has more than one digit; we treat it as 10 + 3, write 3 as the tens' digit of the sum, and convert 10 tens to 1 hundred by carrying a 1 to the hundreds' place, where it is added to the other digits. Thus, in the hundreds' place, we add not 2 + 9 but 1 + 2 + 9, yielding 12. Once again we record 2 in the answer and carry 1 to the next column; 1 + 4 + 3 = 8, the last digit of the sum, which is 8 236.

Subtraction

Subtraction likewise proceeds from right to left. A technique known as borrowing reverses the carrying of addition: when necessary, 1 from the column to the left is converted to 10 in the present column.

$$\begin{array}{r} 8236 \\ 3965 \\ \hline 4271 \end{array}$$

Here we subtract the *subtrahend* 3 965 from the *minuend* 8 236 to yield the *difference* 4 271. As with addition, the digits are aligned by place value. The units' place is easy: 6 − 5 = 1. The tens' place, however, presents a complication: since 3 is less than 6, we cannot subtract 3 − 6 (except by using negative numbers). We borrow 10 by reducing the next digit of the minuend (2) by 1 and adding 10 to the 3 in the minuend; this gives us 13 − 6, which is 7. The hundreds' place then has not 2 − 9 but 1 − 9, since borrowing decreased the minuend.[3] Because 1 is less than 9, we must borrow again: we reduce the 8 in the next column to 7 and add 10

3 Instead of subtracting 1 from the digit in the minuend, one can add 1 to the digit in the subtrahend. This technique is widely used in Europe; it used to be taught in North America as well.

×	2	3	4	5	6	7	8	9	10	11	12
2	4	6	8	10	12	14	16	18	20	22	24
3	6	9	12	15	18	21	24	27	30	33	36
4	8	12	16	20	24	28	32	36	40	44	48
5	10	15	20	25	30	35	40	45	50	55	60
6	12	18	24	30	36	42	48	54	60	66	72
7	14	21	28	35	42	49	56	63	70	77	84
8	16	24	32	40	48	56	64	72	80	88	96
9	18	27	36	45	54	63	72	81	90	99	108
10	20	30	40	50	60	70	80	90	100	110	120
11	22	33	44	55	66	77	88	99	110	121	132
12	24	36	48	60	72	84	96	108	120	132	144

to the 1 in the present column, thereby obtaining $11 - 9 = 2$. Finally, $7 - 3 = 4$, and the answer is 4 271.

Multiplication

Multiplication requires a thorough knowledge of the multiplication table, at least up to 9×9. Here the full table up to 12×12, as it used to be taught, is presented. Omitted are the products involving factors of 0 (any number times 0 is 0) and 1 (multiplying by 1 leaves a number unchanged).

The facts for 10 are superfluous, since multiplying an integer by 10 simply appends a 0. Likewise, 11 times any digit is just that digit repeated, so the facts for 11 (other than $11 \cdot 11 = 121$ and $11 \cdot 12 = 132$) need not be memorised.

Because multiplication is commutative, this table is symmetrical: each column is identical to the corresponding row, and the triangular sections above and below the diagonal containing the squares ($2 \cdot 2, 3 \cdot 3, \ldots$) are mirror images. For that reason, there is no need to learn the entire table: if you know $7 \cdot 4$, you know $4 \cdot 7$. Chinese schools often omit the duplicate entries, teaching only $2 \cdot 2$ to $2 \cdot 9$, $3 \cdot 3$ to $3 \cdot 9$, $4 \cdot 4$ to $4 \cdot 9$, and so on. Nevertheless, you must know the table so thoroughly that you can automatically give any product on it, without thinking.

To multiply a number of multiple digits by a single digit, we proceed from right to left, carrying any excess as in addition.

$$\begin{array}{r} 469 \\ 7 \\ \hline 3283 \end{array}$$

Start with $9 \cdot 7 = 63$; put the 3 down as the rightmost digit of the answer and carry the 6. Then $6 \cdot 7 = 42$, and adding the carried 6 yields 48; write 8 and carry 4. Finally, $4 \cdot 7 = 28$, plus 4 carried gives 32, the rest of the answer. Thus $469 \cdot 7 = 3\,283$.

Multiplication of larger numbers takes advantage of the distributive property of multiplication over addition. In the following example, $469 \cdot 251$, we change one difficult multiplication problem into the sum of three simple ones, each having a one-digit factor: $469(1 + 50 + 200) = 469 \cdot 1 + 469 \cdot 50 + 469 \cdot 200$.

$$\begin{array}{r} 469 \\ 251 \\ \hline 469 \\ 2345 \\ 938 \\ \hline 117719 \end{array}$$

Each partial product is written in turn; those with final zeroes are indented (there is no need to write the zeroes). The three partial products are then added to yield 117 719.

Multiplying with decimal places is easy: simply perform the calculation without regard for decimal points; the product will then have as many decimal places as the factors combined, and the decimal point can be inserted at the appropriate place.

Division

Division differs from the previous operations in two important respects. First, calculation proceeds from left to right. Second, division of one integer (the *dividend*) by another (the *divisor*) does not always yield an integer. For example, to divide 17 items into 3 groups of equal size, we can put 5 into each group, but 2 will remain undistributed. We say that the *quotient* of 17 ÷ 3 is 5 and that the *remainder* is 2. Instead, we can treat the remainder as a fraction with the divisor as its denominator and say that the quotient is $5\frac{2}{3}$.

The following example uses long division to determine that 20 417 ÷ 38 = 537 with a remainder of 11:

```
        537
38 ) 20417
      190
      ---
      141
      114
      ---
       277
       266
       ---
        11
```

We start by taking just enough digits from the dividend to make a number greater than or equal to the divisor: we need three digits (204) because both 2 and 20 are too small. The first digit of the quotient is the largest digit that, when multiplied by 38, yields a number less than or equal to 204. That digit is 5 because 38 × 5 = 190, which is less than 204, and 38 × 6 = 228, which is too large. We write this digit 5 directly over the last digit of 204. Beneath 204 we write 38 × 5, which is 190, and subtract, to obtain 14. The difference must be less than the divisor, 38; if it is not, the digit that we have chosen for the quotient is too small.

We then bring down the next digit of the dividend (a 1) by writing it after the difference of 14, to yield 141. Next, we repeat the process by finding the digit that, when multiplied by 38, comes as close as possible to 141 without going over. Although we might guess 4, it is too high: $38 \times 4 = 152 > 141$. The correct digit is 3. We write the 3 as the next digit of the answer, then write $38 \times 3 = 114$ beneath 141 and subtract, obtaining 27.

We bring down the final digit, a 7, to make 277, and find that the next digit of the answer is 7: $38 \times 7 = 266$. We write 7 as the last digit of the answer and subtract 266 from 277 to obtain the remainder, 11. The answer, then, is 537 remainder 11 or, in fractional form, $537\frac{11}{38}$ (the divisor becomes the denominator of the fraction).

We could instead continue the division by writing a decimal point and calculating as many decimal places as we pleased:

```
         537.289
     ___________
38 ) 20417.000
     190
     ---
      141
      114
      ---
       277
       266
       ---
        110
         76
        ---
         340
         304
         ---
          360
          342
          ---
           18
```

If the divisor contains a decimal point, we convert it to an integer by moving the decimal points in both the dividend and the divisor forward. For example, we change $20417 \div 0.38$ to $2041700 \div 38$ by moving both decimal points two places to the right.

Divisibility and prime numbers

A number a is *divisible* by a number b if $a \div b$ has no remainder (a remainder of 0). For instance, 12 is divisible by 2, 3, and 4 but not by 5.

A *prime number* is a positive integer that is not divisible by any other integer than itself and 1. A positive integer that is not prime is *composite*. (By convention, 1 is neither prime nor composite.)

The first few prime numbers are 2, 3, 5, 7, 11, 13, and 17. The others between 2 and 17 are all composite because they can be factored into primes: for example, $4 = 2 \cdot 2$, $12 = 2 \cdot 2 \cdot 3$, and $15 = 3 \cdot 5$.

Any integer greater than 1 can be factored into primes in only one way (disregarding the order of the prime factors).

Fractions

Any fraction can be expressed in infinitely many ways. For example, $\frac{1}{2} = \frac{2}{4} = \frac{3}{6} = \frac{-5}{-10}$. These fractions are equivalent because any of them can be obtained by multiplying the numerator and denominator of another by a common factor: $\frac{3}{6} = \frac{3 \cdot 1}{3 \cdot 2} = \frac{3}{3} \cdot \frac{1}{2} = \frac{1}{2}$. Indeed, any common factor k other than 1 may be cancelled out to leave a simpler fraction:

$$\frac{ka}{kb} = \frac{k}{k} \cdot \frac{a}{b} = 1 \cdot \frac{a}{b} = \frac{a}{b}.$$

By cancelling all such common factors, we express the fraction *in lowest terms.* This can be done by factoring the numerator and the denominator into primes and omitting any factors that they have in common:

$$\frac{156}{234} = \frac{\cancel{2} \cdot 2 \cdot \cancel{3} \cdot \cancel{13}}{\cancel{2} \cdot \cancel{3} \cdot 3 \cdot \cancel{13}} = \frac{2}{3}.$$

Sometimes it is easier to test only the prime factors of the smaller number. In the fraction $\frac{6}{7\,021}$, there is no need to factor the large denominator: since $6 = 2 \cdot 3$, it is sufficient to determine whether either 2 or 3 divides the denominator evenly. The answer is no, so the fraction is already in lowest terms. (Note that any fraction whose numerator is 1 is already in lowest terms, since no factor can be divided out of the numerator.)

Addition and subtraction of fractions are easiest when the fractions have a common denominator. To add $\frac{2}{3}$ and $\frac{1}{12}$, it helps to convert the first fraction to twelfths by multiplying top and bottom by 4:

$$\frac{2}{3} + \frac{1}{12} = \frac{2 \cdot 4}{3 \cdot 4} + \frac{1}{12} = \frac{8}{12} + \frac{1}{12} = \frac{8+1}{12} = \frac{9}{12}.$$

The answer can then be reduced to lowest terms: $\frac{9}{12} = \frac{3}{4}$.

Sometimes both fractions must be converted to a different denominator:

$$\frac{3}{7} - \frac{1}{5} = \frac{3 \cdot 5}{7 \cdot 5} - \frac{1 \cdot 7}{5 \cdot 7} = \frac{15}{35} - \frac{7}{35} = \frac{8}{35}.$$

The answer is already in lowest terms.

Multiplication of fractions is easier: simply multiply the numerators and the denominators:

$$\frac{1}{5} \cdot \frac{3}{7} = \frac{1 \cdot 3}{5 \cdot 7} = \frac{3}{35}.$$

Division of fractions works just like multiplication except that the reciprocal of the divisor is used:

$$\frac{1}{5} \div \frac{3}{7} = \frac{1}{5} \cdot \frac{7}{3} = \frac{1 \cdot 7}{5 \cdot 3} = \frac{7}{15}.$$

Percentages are just fractions with a denominator of 100. For instance, 75% is $\frac{75}{100}$, or in lowest terms $\frac{3}{4}$. Expressed as a decimal, $75\% = 0.75$.

Powers

Raising a number b (the *base*) to a positive integral *power* n, also called *exponentiation*, means multiplying n factors of b. Thus 2^4 («two to the fourth (power)») is $2 \cdot 2 \cdot 2 \cdot 2$, or 16.

When the base is 1, any power of it is also 1. When the base is greater than 1, raising it to higher and higher powers results in rapid (*exponential*) growth, without limit; the powers approach infinity. When the base is positive but less than 1, higher and higher powers become smaller and smaller, approaching 0. For instance, $0.99^{1\,000}$ is approximately 0.000 043.

Raising any number to the power 1 yields the number itself. Raising any number other than 0 to the power 0 yields 1 (not 0!).[4]

The second power is commonly called the *square*; the third power is commonly called the *cube*.

4 For different purposes, mathematicians either treat 0^0 as 1 or state that it is undefined. In algebra, $0^0 = 1$ is the only sensible definition within the system of real numbers; within the complex numbers, 0^0 is undefined because 0^z is undefined for *any* number z, for reasons to do with the logarithmic definition of complex exponentiation. In the calculus and real analysis, 0^0 often appears not as a static numerical value but as the result of a limiting process that gives the value—if any—of x^y as x and y both become closer and closer to 0. The limit of x^y can be *any* non-negative number, or no number at all, depending on what x and y represent. Thus 0^0 *as the value of a limit* is undefined; we say that it is *indeterminate*.

Remember the following important rules of powers.

$$b^{m+n} = b^m b^n;$$
$$(b^m)^n = b^{mn};$$
$$(ab)^n = a^n b^n.$$

These follow from the definition of a power and from the commutative property of multiplication.[5] For example,

$$b^{m+n} = \underbrace{\underbrace{bbb\cdots b}_{m \text{ times}}\,\underbrace{bbb\cdots b}_{n \text{ times}}}_{m+n \text{ times}}$$
$$= b^m b^n.$$

Negative powers are reciprocals of positive powers. For example, $x^{-1} = 1/x$, and $x^{-3} = 1/x^3$. This definition is consistent with the rules of powers: $b^0 = b^{1-1} = b^1 b^{-1} = b \cdot \frac{1}{b} = 1$, as expected.

Powers that are not integers are discussed below.

Roots

Unlike addition and multiplication, exponentiation is not commutative: in general, x^n is not equal to n^x. For example, $2^3 = 8$, but $3^2 = 9$.

Exponentiation therefore has two inverse operations. One is the extraction of the nth *root*: we seek a number that when raised to the power n yields a given number y. The nth root of y is written $\sqrt[n]{y}$, and when $x^n = y$, we have $x = \sqrt[n]{y}$.

5 Only rational powers are defined in terms of repeated multiplication; irrational powers have a different definition, which appears in the section on exponentials. Nevertheless, these rules for powers apply to *all* powers.

The *square root* of y, the number whose square is y, could be written $\sqrt[2]{y}$ but is usually written $\sqrt{y}$. It is by far the most common of the roots. The *cube root* of y is $\sqrt[3]{y}$.

The nth root of y is a fractional power: $\sqrt[n]{y} = y^{1/n}$. This is because the nth root to the nth power is y. By the rules of powers, $(y^{1/n})^n = y^1 = y$. Therefore, $y^{1/n} = \sqrt[n]{y}$.

A positive number actually has two square roots, each the negative of the other. For instance, $\sqrt{9} = 3$, because $3^2 = 9$; but $(-3)^2$ is also 9, so -3 is another square root of 9. For convenience' sake, $\sqrt{y}$ refers to the *principal square root* of y, the one that is non-negative.

What is the square root of a negative number—say, $\sqrt{-1}$? No real number has a negative square: positive times positive is positive, negative times negative is also positive, and zero times zero is zero. Mathematicians therefore long rejected $\sqrt{-1}$ as a meaningless expression; at best, they considered it «imaginary».[6] They

6 If you are inclined to this view, consider that other types of numbers, too, may seem meaningless or absurd in the wrong context. Humans invented numbers for counting things such as apples in a basket—a purpose for which the small positive integers were good enough. For counting, fractions are useless: one cannot have $\frac{2}{3}$ apples in a basket. But one can have $\frac{2}{3}$ of an amphora of wine, and $\frac{2}{3}$ of the apples in a basket may be yellow, and my basket may contain $\frac{2}{3}$ as many apples as yours. Thus fractions do find application to more sophisticated mathematical purposes. Negative numbers are likewise useless for counting objects, which is why mathematicians were slow to adopt them; yet they are useful for accounting. Similarly, the fact that one cannot have $\sqrt{-1}$ apples in a basket or find $\sqrt{-1}$ on one's bank statement does not mean that the imaginary numbers, and the complex numbers to which they give rise, are silly inventions of mathematicians in an ivory tower; they are vitally important in advanced mathematics not used by most of the public, such as the representation of circular or wave-like motion and the solution of some kinds of equations. People whose mathematical needs do not go far beyond ordinary arithmetic may never have an occasion to use, or even to appreciate, complex numbers, just as those who merely count the apples in a basket will not use negative numbers or fractions.

yielded only when the imaginary numbers proved to be essential for solving certain types of equations.

Today we extend the real numbers by including the *imaginary unit*, known as i,[7] which is defined to be a number such that $i^2 = -1$.[8] Thus one square root of -1 is i. With i as another number, we can form infinitely many others, such as $3 + 4i$ and $\sqrt{7} - \frac{3}{5}i$. In general, those numbers of the form $a + bi$, for any real numbers a and b, are called the *complex numbers*, and they are of immense importance both to advanced mathematics and to modern life: they turn up in such varied settings as electrical circuits, compression of data, computer graphics, forecasting of weather, and the motions of subatomic particles.

The fundamental theorem of algebra implies that any real number (and indeed any complex number) has two square roots, three cube roots, four fourth roots, and in general n nth roots. One or two of the nth roots of a positive radicand are themselves real numbers; the others are all complex numbers. For example, it is easy to see that 1 is a cube root of 1; what is not so easy to see is that the other two cube roots are

$$-\frac{1}{2} + \frac{\sqrt{3}}{2}i \qquad \text{and} \qquad -\frac{1}{2} - \frac{\sqrt{3}}{2}i.$$

This book does not present calculations involving complex numbers; however, it is well to understand that these numbers do exist and that they are of great importance to mathematics, science, engineering, and technology.

Roots can be calculated digit by digit in a manner akin to long division. Here the traditional algorithm for the square root is presented; similar algorithms for cube roots and higher roots

7 Because electrical engineers use i for current, they substitute j for the imaginary unit.

8 Yes, *a* number, because it is not the only one: since negative times negative is positive, it is also true that $(-i)^2 = -1$. The square roots of -1 are i and $-i$.

exist, but they are so tedious that they are seldom taught or used.[9] (Roots other than the square root are more easily calculated by means of logarithms, which are discussed in a subsequent section of this chapter.)

	1	5	3.	1	5
√	2	34	56.	78	
	1				
25	1	34			
	1	25			
303		9	56		
		9	09		
3061			47	78	
			30	61	
30625			17	17	00
			15	31	25

Begin by grouping the digits of the *radicand* into pairs on each side of the decimal point. (As in this example, the leftmost position may have only a single digit. It is a «pair» whose first digit happens to be 0.) Each pair will correspond to a single digit in the answer, which will be written above the *radical sign* ($\sqrt{}$).

The first digit of the root is the number whose square comes the closest to the first pair of digits without exceeding it. Write that digit in the answer, then write its square beneath the first pair of digits and subtract. Bring down the next pair of digits. In the example, we need a digit whose square is exactly or nearly 2; that digit is 1. We write 1 in the answer (above the 2) and subtract its square ($1^2 = 1$) from 2, getting 1. We then bring down

9 The algorithm for cube roots can be found in many arithmetic textbooks from the early 1900s. A nineteenth-century entrance exam for Harvard University called for manual calculation of a cube root.

the next pair of digits, 34, to obtain 134 for the next step of the calculation.

For the next digit, begin by doubling the root that we have so far and writing the result to the left, taking care to leave space after it for an additional digit. The root so far is 1; double that is 2, so we write 2, with space after it, to the left of 134. The next digit of the root will be written both in the answer and in the space that we left after the 2 (the space here is shown by an underscore); thus the 2 will become a number between 20 and 29, inclusive. That new number, when multiplied by the new digit of the root, must come as close as possible to 134 without going over. Finding the new digit may require trial and error.

A good starting point is to append a 0 to the doubled amount (2) and divide the resulting number into 134. Since $134 \div 20$ is 6 (ignoring the remainder), we might suppose that 6 is the new digit of the root. But $26 \cdot 6 = 156$, which is greater than 134. We try 5 instead: $25 \cdot 5 = 125$, which is no greater than 134, so indeed 5 is the next digit. Write it in the answer and also after the 2; then multiply, subtract, and bring down another pair of digits. We have already calculated 125 as the product; subtracting $134 - 125$ yields 9, which becomes 956 when we bring 56 down.

Repeat the process for each subsequent digit. The root so far is 15; double that is 30, after which we leave a space. The next digit of the answer is 3 because $303 \cdot 3 = 909$ brings us as close as we can get to 956. Record the 3, subtract $956 - 909$ to obtain 47, and bring 78 down to obtain 4778.

After writing the decimal point in the root, we double 153, yielding 306. The next digit can only be 1: $3061 \cdot 1 = 3061$, which, subtracted from 4778, yields 1717, to which we append 00 as the next two (hidden) digits of the radicand.

We can continue as long as we wish, but let us stop after one more digit. Doubling yields 3062 (disregard the decimal point); the next digit is 5; the product is $30625 \cdot 5 = 153125$. Since we end the calculation here, we need not perform the subtraction.

(We could also have avoided the multiplication by noticing that 5 was certainly the correct digit, neither too small nor too large, because the product would start with 15 and therefore computing it was unnecessary.) Thus $\sqrt{23\,456.78} \approx 153.15$.

Although some roots do come out exact (such as $\sqrt{49} = 7$), most are irrational numbers: their digits go on forever, but without the repeating patterns that rational numbers form. Irrational numbers can only be approximated; they cannot be represented exactly, even by fractions (which are rational).

Logarithms

The other inverse operation for exponentiation is the *logarithm*. Whereas a root gives us the base that must be raised to a given power in order to yield a given number, the logarithm gives us the power to which a given base must be raised to yield the number in question. The logarithm to the base b of a number y is written $\log_b y$. Examples:

$$\begin{aligned} \log_2 8 &= 3, &\text{because}\quad 2^3 &= 8; \\ \log_2 32 &= 5, &\text{because}\quad 2^5 &= 32; \\ \log_{10} 1\,000\,000 &= 6, &\text{because}\quad 10^6 &= 1\,000\,000; \\ \log_7 \tfrac{1}{7} &= -1, &\text{because}\quad 7^{-1} &= \tfrac{1}{7}. \\ \log_9 3 &= \tfrac{1}{2}, &\text{because}\quad 9^{1/2} &= \sqrt{9} = 3. \end{aligned}$$

The base b must be a positive number other than 1, and the number whose logarithm is sought must also be positive.[10] These facts are true of all bases:

$$\begin{aligned} \log_b 1 &= 0, &\text{because}\quad b^0 &= 1; \\ \log_b b &= 1, &\text{because}\quad b^1 &= b. \end{aligned}$$

10 Negative numbers have complex logarithms, with infinitely many values. Zero does not have a logarithm at all.

Two logarithms are of particular importance. The *common logarithm* uses base 10. It is especially useful for mental calculation because its base is that of our decimal number system. Because of its importance, it is often written simply as $\log y$ rather than as $\log_{10} y$.

The *natural logarithm* has as its base a vitally important irrational number called e, approximately 2.718 281 828 459. It is «natural» because of the manner in which it is defined in the integral calculus.[11] Usually it is written $\ln y$, but mathematicians often treat it as *the* logarithm and write it as $\log y$ instead. On a calculator, 'log' refers to the common logarithm and 'ln' to the natural logarithm.

Logarithms were introduced to simplify calculations. The following properties, which can be derived from the rules governing powers, enable them to transform difficult calculations into simpler ones. (Here 'log' means the logarithm to any base, not just the common logarithm.)

$$\begin{aligned}\log xy &= \log x + \log y;\\ \log \frac{x}{y} &= \log x - \log y;\\ \log x^y &= y \log x;\\ \log \sqrt[y]{x} &= \frac{1}{y} \log x.\end{aligned}$$

Logarithms can thus convert multiplication to addition, division to subtraction, exponentiation to multiplication, and the extraction of roots to division. In each case, the task is much simpler. The great French mathematician Pierre Simon de Laplace praised

11 Specifically, $\ln x = \int_1^x (1/t)\, dt$. Whether or not you understand this notation, you can appreciate its simplicity, which accounts for its representing the *natural* logarithm. The logarithm to any other base is the same expression multiplied by a number, typically irrational, that would appear arbitrary, messy, and anything but natural.

> la découverte des logarithmes, ... artifice admirable ... qui en réduisant à quelques jours, le travail de plusieurs mois, double, si l'on peut ainsi dire, la vie des astronomes, et leur épargne les erreurs et les dégoûts inséparables des longs calculs ...
>
> the discovery of logarithms, an admirable invention that, by reducing to a few days the work of several months, doubles, if you will, the lives of astronomers and spares them the errors and unpleasantness that necessarily accompany lengthy calculations.[12]

Because logarithms are generally difficult to calculate one by one, astronomers and others over the centuries looked them up in published tables.[13] For common logarithms, which were used for simplifying calculations, a «log table» gives values for numbers in the range from 1 to 10. Since these logarithms are all between 0 and 1, their integral part—to the left of the decimal point—is 0; it is known as the *characteristic*. The table gives only the *mantissa*—the decimal portion of the logarithm. Numbers outside the range are factored to obtain an integral power of 10 times a number in the range; the power is then the characteristic.

If $x \geq 1$, the characteristic of $\log x$ is 1 less than the number of digits in x to the left of the decimal point; if $0 < x < 1$, the characteristic is negative, with a value 1 greater than the number of zeroes just to the right of the decimal point. For example, log 3.74 can be found directly in a table, because $1 \leq 3.74 < 10$. If the table gives logarithms to five decimal places, as many good tables do, the mantissa will appear in the table as 57287; thus $\log 3.74 = 0.57287$. Although log 374 does not appear in the table, we note that its characteristic is 2 (1 fewer than the number

12 [Pierre Simon de] Laplace, *Exposition du système du monde* (Paris: Courcier, 1808) at 368. My translation.

13 These tables are obsolete today, but I still used them into the mid-1980s.

of its digits); its logarithm is therefore $2 + \log 3.74 = 2.57287$. (Using the rules above, $\log 374 = \log(100 \cdot 3.74) = \log 10^2 + \log 3.74 = 2 + 0.57287 = 2.57287$.)

To get log 0.0374, we note that the characteristic is −2, because there is 1 zero just after the decimal point and the characteristic is the negative of the number 1 greater than that. The answer, however, is not −2.57287: it is $-2 + 0.57287$, which is −1.42713. Because the table gives only positive mantissas, however, it is customary during calculations to preserve the mantissa from the table and mark the characteristic alone as negative, either by writing a bar over it ($\bar{2}.57287$) or by adding 10 to it and writing « − 10» at the end ($8.57287 - 10$).

The properties of logarithms listed above enable us to simplify calculations. Suppose that we want $392.17 \cdot 2.1886$. Using a log table, we find that $\log 392.17 = 2.59347$ and $\log 2.1886 = 0.34017$. Their sum, 2.93364, is the *logarithm* of the product. To obtain the product itself, we use the log table in reverse to determine that the *antilogarithm* of 2.93364 is 858.30.[14] Indeed, $392.17 \cdot 2.1886 = 858.30$ to five decimal places.

To obtain $392.17 \div 2.1886$, we subtract instead: $2.59347 - 0.34017 = 2.25330$. The antilogarithm, 179.18, is close to the correct answer, 179.19. (With five-digit logarithms, we cannot expect more than four correct significant digits; the fifth digit may be off slightly.) For a power such as 392.17^2, we multiply the logarithm of the base by the exponent: $2.59347 \cdot 2 = 5.18694$; taking the antilogarithm yields 153 790 to five digits (correct answer to the same accuracy: 153 800; the discrepancy is due to a slight difference in rounding in the *sixth* digit). For a root, such as $\sqrt[3]{392.17}$, we divide the logarithm of the base by the degree of

14 We determine the antilogarithm by finding the mantissa, 93364, in the table and seeing that it (rather, 0.93364) is the logarithm of 8.5830. Then we adjust the decimal point to match the characteristic: there must be $2 + 1 = 3$ digits to the left of the decimal point, so the antilogarithm is 858.30.

the root: $2.59347 \div 3 = 0.86449$; the antilogarithm is 7.3196 (correct answer: 7.3197, again because of rounding).

We can convert a logarithm from one base to another simply by dividing by the logarithm of the new base:

$$\log_b x = \frac{\log_a x}{\log_a b}.$$

Thanks to this formula, we can use the common logarithm exclusively for calculation. When we need the natural logarithm, we can calculate the common logarithm instead and convert it in this simple way:

$$\begin{aligned}\ln x = \log_e x &= \frac{\log x}{\log e} \\ &\approx \frac{\log x}{0.43429} \\ &\approx 2.3026 \log x.\end{aligned}$$

For instance, if we know that $\log 2 = 0.30103$, we can obtain $\ln 2 = 2.3026 \log 2 = 2.3026 \cdot 0.30103 = 0.69315$.

Exponentials

Integral and fractional powers are discussed above. What about irrational powers? Can one speak of $2^{\sqrt{3}}$, or of 10^e?

For mental calculation, these could simply be approximated as the rational powers $2^{1.7321}$ and $10^{2.7183}$, perhaps with fewer digits still. The answer to the question, however, is yes: these expressions have an exact meaning that is based on the natural logarithm and its inverse, the *exponential function*. If $x = \ln y$, then the exponential $\exp x$, or more simply e^x, is y. Thus the exponential function and the natural logarithm are inverses, each undoing the other: $e^{\ln x} = x$ and $\ln e^x = x$. In other words, the exponential function is the antilogarithm to the base e.

The definition of e^x gives us a way to calculate *any* power x^y of a positive number x:[15]

$$x^y = e^{\ln x^y}$$
$$= e^{y \ln x}.$$

In practice, we use base 10 instead of base e for calculation:

$$x^y = 10^{y \log x}.$$

Just as e^x is the inverse of $\ln x$, so 10^x is the inverse, or *antilogarithm*, of $\log x$. The antilogarithms of x found in the previous section by using a log table in reverse are simply 10^x.

Both e^x and 10^x are difficult to compute; people over the centuries have relied variously on tables, slide rules, calculators, or computers. A good technique for computing e^x follows:

$$e^x = 1 + x + \frac{x^2}{2} + \frac{x^3}{2 \cdot 3} + \frac{x^4}{2 \cdot 3 \cdot 4} + \frac{x^5}{2 \cdot 3 \cdot 4 \cdot 5} + \cdots$$

This equation works best for small values of x.[16] Forbidding though it may seem, it is actually easy to use in calculation: each term is simply the preceding term multiplied by $\frac{x}{n}$ (where n is the power of x in the term). We can thus take advantage of calculations already performed and spare ourselves a great deal of work.

Because $10^x = e^{x \ln 10}$, the same equation can be used to compute 10^x. In practice, however, e^x, 10^x, and x^y for awkward values of y are too cumbersome for manual calculation—which is why *mental* calculation of them is so impressive.

15 Although y can be any number, x must be positive, because the definition of x^y that is given here employs $\ln x$, which exists (in the real numbers) only for positive values of x. Where y is an integer, however, x^y for non-positive values of x retains the definition given in the section on powers. Arbitrary powers of negative numbers exist as complex numbers, and in general those powers have multiple values. The definition of complex exponentiation, however, goes beyond the scope of this book and is therefore omitted.

16 It can be used to calculate e itself: $e = e^1 = 1 + 1 + \frac{1}{2} + \frac{1}{2 \cdot 3} + \frac{1}{2 \cdot 3 \cdot 4} + \cdots$. The first 13 terms, easily computed on paper, yield e to 10 digits' accuracy. Try it!

Trigonometric functions

Trigonometry is rooted in the measurement of the sides and angles of triangles, although it has applications to countless fields whose relationship to triangles is not obvious, among them electrical engineering, quantum mechanics, digital audio and video, medical imaging, the solution of equations, even logarithmic and exponential functions.

A full course in trigonometry lies far beyond this book's purpose; please consult instead any good textbook.[17] This section merely introduces the main trigonometric functions and their inverses.

Consider the right triangle below:

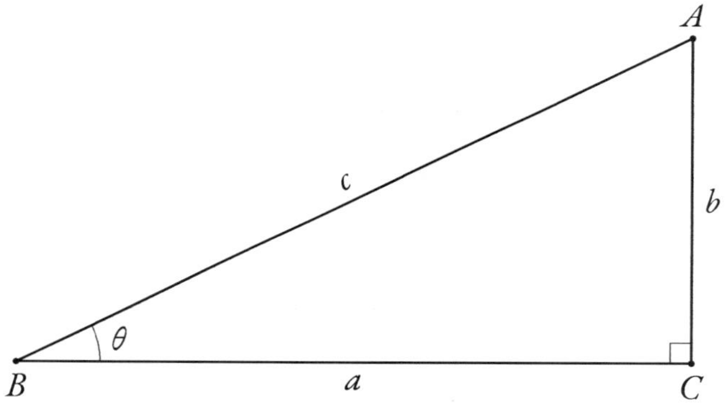

The side *opposite* an angle is that side that is not part of the angle. By convention, an angle is labelled with a capital letter; its opposite side is labelled with the corresponding small letter. The side opposite the right angle (angle C in the diagram) is called

17 The great Indian mathematician Srinivasa Ramanujan, who flourished at the turn of the twentieth century only to succumb to amœbiasis at the frightfully young age of 32, cut his teeth on Sidney Luxton Loney's *Plane Trigonometry* (Cambridge: University Press, 1895). You could do worse than to download this textbook from the Internet Archive at ⟨https://archive.org⟩.

the *hypotenuse*. In an angle that includes the hypotenuse, the remaining side is called the *adjacent*.

Angle B is marked as having angular measure θ (theta). Its opposite (opp) has length b; its adjacent (adj) has length a; the hypotenuse (hyp) has length c.

Now we can define the three most important trigonometric ratios. For an angle θ in a right triangle, the *sine* ($\sin\theta$), *cosine* ($\cos\theta$), and *tangent* ($\tan\theta$) are defined as follows:

$$\sin\theta = \frac{\text{opp}}{\text{hyp}} = \frac{b}{c};$$
$$\cos\theta = \frac{\text{adj}}{\text{hyp}} = \frac{a}{c};$$
$$\tan\theta = \frac{\text{opp}}{\text{adj}} = \frac{b}{a}.$$

These ratios are well defined for every angle because all triangles with the same angles are similar, which means that they differ only in size: every triangle with the angles of the one shown above has sides that measure ka, kb, and kc, for some scaling factor k. In any such triangle, $\sin\theta = \cancel{k}b/\cancel{k}c = b/c$; the scaling factor falls out, leaving a consistent value for the sine irrespective of the right triangle that is used to calculate it. The same is true for the cosine and the tangent.

Calculating trigonometric functions is difficult for all but a few angles. Students of the calculus learn ways to compute these functions to any desired degree of accuracy, but the techniques are too laborious for manual calculation. Instead, people ordinarily rely on the values found in tables or, nowadays, on calculators. Nevertheless, this book does present ways to approximate these functions mentally.

The triangular model used in this section allows for defining the sine, cosine, and tangent of any angle greater than 0° and less than 90°. A different model, based on the circle, extends the definitions of these functions to all angles—positive, negative, or zero.[18]

Each trigonometric function has an inverse, whose name begins with «arc-». The *arcsine* of x ($\arcsin x$) is the angle whose sine is x.[19] The *arccosine* ($\arccos x$) and *arctangent* ($\arctan x$) are defined in like manner.

People commonly measure angles in degrees, with 90° in a right angle. Mathematicians, however, usually measure angles in *radians*: a right angle measures $\frac{\pi}{2}$ radians, in which π (pi) is the ratio of the circumference of a circle to its diameter, approximately 3.141 592 653 589 793. One degree is therefore $\frac{\pi}{180}$ radians (about 0.0175); one radian is $\frac{180}{\pi}$ degrees (about 57.3).

18 The tangent, however, does not exist at 90° plus any multiple of 180°.

19 Under the circular definition of the trigonometric functions, infinitely many angles have the same sine, cosine, or tangent. The arcsine, arccosine, and arctangent are each defined such that their values lie within a given range and are unique.

Chapter 2

Estimation

MASTERY of the material in this book will enable you to perform in your head such challenging calculations as muliplication yielding products of ten or twelve digits, extraction of square roots to six or seven digits' accuracy, logarithms to five decimal places. But why calculate more than you need to? When a few digits are good enough, make do with an estimate.

Before electronic calculators became prevalent, estimation saved a lot of laborious and unnecessary numerical work. Now that ubiquitous calculators will readily generate eight or more digits, estimation is not exploited to greatest advantage: people who could easily estimate a calculation in their heads will instead reach for a calculator and obtain an «exact» answer with more precision than they need. Reviving the art of estimation will serve you well in daily life and also enable you to perform mental arithmetic that can only be done approximately.

Significant figures

Evaluate the following four statements before reading further.

> The population of Canada is 40 884 967.
>
> Here one can retire in comfort on as little as $1 743 per month.

> The nearest grocery store is 6 miles (9.656064 kilometres) away.
>
> The Plaintiff seeks general damages in the estimated amount of $300 000.00.

Similar statements appear every day in the media, even in academic publications. They are wrong because they promise more accuracy than they can deliver.

The first statement implies that the population of Canada is known exactly, and that it cannot be a single person higher or lower than the figure given. Yet the population could not be measured exactly at any particular moment. If it *were* measured exactly, the statement would not be correct unless it included the date and time—within ten seconds or so—, since births, deaths, and migration cause it to change thousands of times per day. Even the definition of *population* will prove controversial: does it include all temporary residents? people whose house straddles the US border? people who divide their time between Canada and another country?

The population can be measured accurately to only a few digits. It would be better to state «As of 2024, the population of Canada is 40.9 million» or «... is 40.88 million», depending on the accuracy that the demographers claim.

The second statement implies that the budget given is accurate down to the dollar, which is to say within $0.50. It suggests that a retiree could be comfortable with $1 743 per month but feel deprived with only $1 742. That cannot be correct: needs and expenditures fluctuate, as do exchange rates and prices. It would be better to state that a particular retiree spent on average $1 743 per month over the past year, or that one could retire comfortably on $1 800 per month, or that $2 000 per month was more than enough for comfortable retirement.

In the third statement, the figure for miles was mechanically converted without regard for the meaning of those six decimal places. The figure for kilometres is stated down to the millimetre (1 mm = 0.000 001 km); it implies that the distance is known within 0.5 millimetres. Were that so, the distance in miles would be known within 0.02 inches. But «6 miles» implies an error no greater than 0.5 miles (*i.e.*, the true distance is between 5.5 and 6.5 miles). The converted figure should have been rounded to 10 kilometres.

The error in the fourth statement may be subtler and harder to discern. I see it frequently in my work as a lawyer, however; I mention it here in the unrealistic hope that lawyers will read about it and correct their ways.

What is wrong with the figure $300 000.00 is that it is stated exactly, right down to the cent: that is what the decimal point and the following two zeroes imply. The amount, however, is said to be «estimated». Presumably the Plaintiff believes that the amount, to be determined later from the evidence, is *around* $300 000—well above $200 000, well below $400 000. It is thus in the range of $250 000 to $350 000, which, rounded to the nearest hundred thousand, gives $300 000.[20] The claim should have been stated as «$300 000», with no decimal point.[21] By convention, that figure is understood to be accurate to the nearest hundred thousand, not to the nearest dollar, just as «She paid

20 The extremes of this range, each of which contains exactly 5 after the leading digit, might be rounded to $200 000 and $400 000 instead. Indeed, they would be so rounded under the commonly used standard of rounding to an even number. Other standards exist, however, and for that reason I do not propose to address the admittedly important question of rounding in greater detail. Note that a 5 followed by anything other than zeroes is rounded up: $250 001 → $300 000.

21 The period at the end of the sentence, it is true, could be misinterpreted as a decimal point. A pernickety lawyer—might I know of one, by chance?—would recast the sentence to leave the figure unambiguously detached from any period.

$300 000 for her house» allows for the possibility of $312 500 or $278 750.

Significant digits, or *significant figures*, are those that indicate a reliably determined value. Zeroes that are written just to show the place value of other digits are not significant themselves.

All non-zero digits are significant. Zeroes between non-zero digits, as in 9 006, are always significant: they indicate a known (measured, determined) value of 0 in their position. Trailing zeroes are significant if at least one of them follows a decimal point, because a non-significant trailing zero would not be written after a decimal point. Otherwise, leading and trailing zeroes are not significant: they are just an artefact of our way of writing numbers, which uses zeroes to fill vacant positions.[22] Thus the zeroes in 300 000 and 0.008 are not significant, but the four zeroes at the end of 700.00 are significant.

In a number known to be exact, all digits are significant. For example, if there are 5 objects in a box, the quantity 5 is known exactly: it is 5.0000... and therefore has infinitely many significant digits. Likewise, π has infinitely many significant digits: we can obtain as many as we wish, and indeed trillions of them have been calculated.

By contrast, measured amounts can usually be given only to a few significant digits. When measuring volume with a cylinder graduated in increments of 10 millilitres, we can ascertain that the volume lies between, say, 130 ml and 140 ml; we can even estimate *one* extra digit and give the volume as approximately 137 ml. That is the best that we can do; any further digits (137.856 ml)

22 One common way to show that a trailing zero is significant is to write a bar over it: in $60\,0\bar{0}0$, the first four digits are significant but the last one is not. Scientific notation also shows significance: 6.000×10^4 has four significant digits.

would be spurious and meaningless, because even the last digit of 137 is not certain.

It is incorrect, and certainly misleading, to present more significant digits than are known. The population of Canada cannot be given to eight significant digits; quite a few of the digits at the end must be spurious. A distance measured with only one significant digit cannot become seven significant digits when converted to another unit of measurement; it should be rounded to one significant digit.

When measured values are added or subtracted, the last significant digit is in the same place as the highest of the last significant digits of the values in question. Thus 6.7 m plus 0.0089 m is not 6.7089 m, because 6.7 m was measured only to tenths of a metre; it is 6.7 m. Similarly, 6.7 m plus 12.345 m is 19.0 m, and 6.7 m minus 0.3456 m is not 6.3544 m but 6.4 m.

When measured values are multiplied or divided, the number of significant digits is that of the value with the fewest significant digits. Thus 123 · 0.6, if the values are measurements, is not 73.8 but 70, because 0.6 has only one significant digit; 45.67 ÷ 89 is not 0.513 146 1, as an eight-digit calculator would state, but rather 0.51.

In a common logarithm, the mantissa is given to as many significant digits as the number itself. A common antilogarithm is computed to as many significant digits as the mantissa.

To retain accuracy, we round for significant digits only at the end of a calculation, not in the middle.

Examples of estimation

When adding or subtracting, simply round the figures at any convenient place. For example, 19 345 + 37 882 can be estimated as 19 000 + 38 000 = 57 000, or more accurately as 19 300 + 37 900 = 57 200; the actual value is 57 227.

The estimate will be better when one number is rounded up and the other is rounded down; then the errors introduced by rounding will offset each other in part: $371+442 \approx 400+400 = 800$ is close to the actual value, 813. When both numbers are rounded in the same direction, the errors are compounded: for instance, $334+228 = 562$, but rounding yields only $300+200 = 500$, which is off by more than 10%.

Estimating a product can be more difficult because multiplication compounds errors considerably: $76 \cdot 47 = 3\,572$ becomes inaccurate if the figures are rounded to $80 \cdot 50 = 4\,000$. Unless that coarse degree of accuracy is acceptable (as it is for certain purposes), either more digits in the factors should be preserved ($76 \cdot 50 = 3\,800$ yields a better estimate) or one factor should be increased and the other decreased, preferably in proportion to their sizes. For instance, 47 could be increased by 3 to yield the conveniently round factor 50; 76, being larger than 47, might then be decreased by an amount somewhat greater than 3, such as 4, to yield 72. The estimate would become $50 \cdot 72 = 3\,600$. A good mental calculator might instead reduce 76 to 75 and raise 47 to 48 because these, for a reason explained in a subsequent chapter, yield $75 \cdot 48 = 3\,600$ by an easy calculation.

When estimating a quotient, be sure to ascertain the magnitude first—a good practice for mental division in general. For instance, $66\,703 \div 92$ is a three-digit number (thus in the hundreds) because the first digit if the calculation were done by long division would be written above the 7 in the dividend, which is the third digit from the right. One could also see that $66\,703 \div 92$ is approximately $66\,703 \div 100 \approx 667$ or approximately $66\,703 \div 90 = 6\,700 \div 9 \approx 700$ and conclude, coarsely, that the quotient was around 700. For a better estimate, reduce 92 by 2 to 90 for convenient calculation and likewise reduce 667 by a proportional amount (perhaps $7 \cdot 2 = 14$, since 667 is roughly $7 \cdot 92$); then the estimate is $653 \div 9 \approx 72.6$ adjusted to three digits, or 726. To three digits' accuracy, the actual quotient is 725. We have therefore estimated accurately and quickly a quotient that would have been troublesome by long division.

Even a square root can be estimated. Although an entire chapter of this book develops a good technique for estimating any square root to six digits or more with ease that may surprise you, you may enjoy an invigorating foretaste of it at this early stage of your studies: simply find the number midway between a reasonable estimate of the square root and the quotient of the radicand divided by that estimate. For example, to estimate $\sqrt{39}$, you might start with $6 = \sqrt{36}$. A better estimate will lie midway between 6 and $39 \div 6 = 6.5$, which is to say 6.25.[23] This is quite close: to four digits, the actual root is 6.245.[24] If instead you wanted $\sqrt{79}$, you might elect $9 = \sqrt{81}$, compute $79 \div 9 = (81 - 2) \div 9 = 9 - \frac{2}{9} \approx 9 - 0.22 \approx 8.78$, and find the midpoint as $9 - 0.22 \div 2 = 8.89$; the correct root is just that, to three decimal places. Even a poor estimate, such as 1.5 for $\sqrt{2}$, may suffice: between 1.5 and $2 \div 1.5 = 4 \div 3 \approx 1.33$ is 1.415 (calculated as the average, $(1.5 + 1.33) \div 2$); the root to four digits is 1.414. Do you still feel that estimation is passé in this digital age?

---

23 This can be computed as an average in the usual way, by adding the numbers and dividing the sum by 2: $(6 + 6.5) \div 2 = 12.5 \div 2 = 12.25$. When two numbers are so close, however, it is easier to add half of the difference to the smaller number (or, as in the following example, to deduct half of the difference from the larger number): the difference is 0.5, half of which is 0.25, so the average is $6 + 0.25 = 6.25$. Think of this method as splitting the difference.

24 Ordinarily I would not choose an example that cast a technique in an especially favourable light. I did so here because Ron Doerfler, in *Dead Reckoning* (at 3), mentioned being appalled that no engineer in a meeting could estimate $\sqrt{39}$. You can now upstage a team of engineers without even breathing hard. Ron, refer those jokers to this book.

Chapter 3

Addition

So basic is the skill of addition that it offers few opportunities for streamlining; mental arithmetic shows to greater advantage in multiplication and other sophisticated operations. Yet a solid foundation in addition underlies more advanced calculations. This chapter presents proper technique for efficient mental addition, whether of two numbers or a whole column.

Proceed from left to right

The algorithms taught in the first chapter for addition, subtraction, and multiplication operate from right to left so that the written result comes out correct and pristine every time, without erasures or other corrections.

In mental arithmetic, however, we always work *from left to right*. This approach proceeds logically from the most significant digit to the least. It also corresponds to the manner of naming numbers in English and almost all other modern languages.[25]

25 Some languages make slight use of an illogical sequence. English, for instance, has «sixteen», from «six» and «ten», instead of the more logical «ten-six». Larger two-digit numbers used to be represented in this way: we have traces in «Four-and-twenty blackbirds baked in a pie» and (from a poem by A E Housman) «When I was one-and-twenty». Arabic, Dutch, and German use the same reversed pattern for non-multiples of 10 from 21 to 99. Many languages use subtraction or other strategies for certain numbers, such as Latin *duode-*

We find it easier to remember numbers whose digits occur in the sequence that is natural to us.

It is true that we have to alter our calculations as we go because of carrying, borrowing, or similar developments. Nevertheless, as will be seen in the next section, calculating from left to right wins hands down in mental arithmetic.

Adding two numbers

This example appeared in the first chapter:

$$\begin{array}{r} 4271 \\ 3965 \\ \hline 8236 \end{array}$$

Performed from right to left as if on paper, this calculation would generate the digits in reverse order, from least to greatest place value. They would have to be reversed from 6328 to 8236. They would be a mere string of digits, not a *number*. The answer could not be given even in part until the end.

By contrast, proceeding from left to right enables us to remember them in their natural and linguistic order, to think of them as a coherent number (with so many thousands, hundreds, and so forth), to begin to recite the answer long before it is finished, and even to abandon the calculation in the middle and make do with a decent estimate, all without the pointless, error-prone task of reversing the digits.

At the simplest level, we start with $4 + 3 = 7$. Already we know that the answer is somewhat, but not much, more than

triginta 'two from thirty' for 28 and Danish *halvtreds* 'half of (the) third', meaning half of the third twenty (and all of the first two twenties), for 50. Some evidence suggests that the ancient Proto-Indo-European language, an ancestor of English, favoured counting up the line, at least for some numbers. I do not, however, know of any modern language that not exceptionally but systematically arranges the place values in any but descending order.

7 thousand. Had we started from the right, we would know only that it ended in 6—a useless fact on its own.

Next is 2 + 9 = 11. Since 11 consists of more than one digit, carrying occurs. We can think of our two «digits» as needing to be recast: 7 followed by 11 is «seventy-eleven», or eighty-one. In this book, we shall use square brackets to manage numbers that exceed the allotted space and thus require adjustment: [7, 11] = [7 + 1, 1] = [8, 1] = 81. It happens that no figure greater than 1 is ever carried during the addition of two numbers, so you could instead just increase 7 by 1 and drop the leading 1 from 11; however, you should practise thinking of numbers in this way and merging them where the place values overlap. Soon enough you will be able to merge them easily, even if at first the technique seems unfamiliar. It may help to think that 11 cannot fit in the single space allotted to the hundreds' place in the calculation, so its leading 1 is bumped into the thousands' place, which is already occupied by 7; the 1 and the 7 must therefore be combined to make 8, and the partial sum is now 81.

At this point, carrying may turn the 1 into 2, but it will not affect the 8, which is therefore certain. We can already begin to announce the answer, as if we had finished the calculation: «8 thousand ...».

Next, 7 + 6 = 13 gives [8, 1, 13] = [8, 2, 3] = 823, or more efficiently [81, 13] = [82, 3] = 823. (Even more efficiently, [1, 13] = 23. Now that we have announced the 8, we may not have to remember it.) We can continue to announce «... 2 hundred ...».

Finally, 1+5 = 6. No carrying occurs; we simply append 6 to 823, obtaining 8 236 as the answer and concluding by announcing «36».

We could also have stopped early, with an estimate: «It's about 8 100 or 8 200.» That would not have been possible when calculating from right to left.

Practise by writing addition problems down so that you do not have to remember their digits while also calculating the sum. If the four-digit example above taxes your memory, try working with two-digit numbers instead. Feel free to do the work aloud, or even on paper, while you master the technique; try it mentally when you have built up your confidence.

Subtraction as an aid to addition

Especially when adding a number that ends in 8 or 9, we can add a higher round number and subtract the excess. Instead of adding 39 directly, for instance, we can add 40 and subtract 1. Thus $42 + 39 = 42 + 40 - 1 = 82 - 1 = 81$.

The same technique works with a difference of two or more digits. For example, $79 + 86 = 79 + 100 - 14 = 179 - 14 = 165$.

Two digits at a time

With practice, you can add two digits at a time for greater speed. In the example of 4 271 + 3 965, compute $42 + 39 = 81$, then $71 + 65 = 136$. Because the second sum goes beyond two digits, the 1 must be carried: $[81, 136] = [82, 36] = 8\,236$.

Adding columns of numbers

Usually adding a long column of numbers involves pencil and paper, so the work proceeds from right to left.

The secret to adding a column of numbers rapidly is to look for combinations that total 10, such as $5 + 5$, $2 + 8$, and $3 + 4 + 3$. Also useful are combinations that add up to a multiple of 10 (usually 20), such as $9 + 4 + 7$.

Practise adding digits in groups of two. When you become skilled at that, practise with groups of three. Once you can add the digits in a column, try to add two columns simultaneously, entry by entry.

One trick to bear in mind: a multiple of 10 from 20 onward can be added to a following digit by simply appending that digit. This works in English, not necessarily in other languages. For example, when adding 9 + 4 + 7 + 8, you might—before gaining enough experience to add the numbers at a glance—think «9 plus 4 is 13, plus 7 is 20». The next digit, however, can be added just by appending it to 20; therefore, instead of thinking «20 plus 8 is 28», think «20... 8», which is to say «twenty-eight», and move on. Use this time-saving trick whenever you reach a total that is a multiple of 10.[26]

```
 493
 271
 655
 149
 304
 868
 943
 467
 512
 536
 270
 926
————
6394
```

In the example above, adding from bottom to top in the first column, we would proceed as follows if adding two digits at a time. Focus on the running total.

«6 plus 6 (skipping the 0) is 12, plus 9 is 21, plus 11 is 32, plus 13 is 45, plus 9 (5 + 1 + 3) is 54. Put down the 4 and carry the 5. Then 5 plus 9 is 14, plus 4 is 18, plus 10 is 28, plus 10 is 38,

26 Except 10 itself: we irregularly say not «ten-eight» but «eighteen».

plus 12 is 50... 9 (59). Put down the 9 and carry the 5. Finally, 5 plus 11 is 16, plus 10 is 26, plus 13 is 39, plus 11 is 50... 7, plus 6 is 63.»

If adding three digits at a time, we proceed as follows, from top to bottom this time (obviously we can add in either direction):

«9 plus 21 is 30, plus 12 is 42, plus 12 is 54. Then 5 carried plus 21 is 26, plus 10 is 36, plus 11 is 47, plus 12 is 59. Finally, 5 carried plus 12 is 17, plus 12 is 29, plus 18 is 47, plus 16 is 63.»

If adding the two rightmost columns at a time from bottom to top:

«26 plus 70 is 96, plus 36 is 132, plus 12 is 144, plus 67 is 211, plus 43 is 254, plus 68 is 322, plus 04 is 326, plus 49 is 375, plus 55 is 430, plus 71 is 501, plus 93 is 594.»

The examples above are written out in full to illustrate the way of proceeding, but you should think simply of the running total: not «x plus y is z» but simply «z».

With practice, you can improve your skill at adding columns of numbers quickly. Put accuracy ahead of speed. When you can add two digits at a time confidently, try looking ahead for larger groups that you can add as a unit, especially those whose total is a multiple of 10. Also take advantage of the facts that $9 = 10 - 1$, $8 = 10 - 2$, $7 = 10 - 3$, and $6 = 10 - 4$. For example, $74 + 9 = 83$: just increase the 7 by 1 and decrease the 4 by 1. Similarly, $74 + 7 = 81$: increase the 7 by 1, decrease the 4 by 3.

Chapter 4

Subtraction

DIFFERENCES somehow seem more difficult than sums, mainly because of the awkward practice of borrowing. By learning a few techniques from mental arithmetic, you will find that subtraction is almost as easy as addition.

Proceed from left to right

Even though figures that have already been calculated may have to be reduced by 1, proceeding from left to right is superior for mental arithmetic. Consider this example from the first chapter:

$$\begin{array}{r} 8236 \\ 3965 \\ \hline 4271 \end{array}$$

Starting at the left, we calculate $8 - 3 = 5$; thus we know immediately that the answer is approximately 5 000. Working from right to left in the manner of manual calculation, we would not know that important fact until the last step.

Then we see that we cannot calculate $2 - 9$, so we reduce the previous result of 5 by 1, yielding 4, and calculate $12 - 9 = 3$. This is our simplified form of borrowing in left-to-right mental arithmetic. Note how much easier it is than the right-to-left procedure used for manual calculation.

The answer so far is 43. Indeed, we can announce «4 thousand ...» because any further borrowing will not affect the 4.

Once again, 3 – 6 forces us to reduce the 3 of the answer to 2; then we have 13 – 6 = 7 and can announce «... 2 hundred ...». Finally, 6 – 5 = 1, and we can announce «... 71» as the rest of the answer. In full, the answer is 4 271.

As this example illustrates, borrowing becomes a mere subtraction of 1 from the most recently calculated digits. By looking ahead, we can anticipate the need for borrowing and work it into our calculation. In the example, seeing that the second position (2–9) required borrowing, we could have calculated 8–4 instead of 8 – 3 and proceeded with 12 – 9.

As long as the last digit calculated is not 0, the digits before it are certain; only if the last digit is 0 might the previous digits be affected by borrowing. Consider this variation on the example:

$$\begin{array}{r} 8236 \\ 4271 \\ \hline 3965 \end{array}$$

By calculating 8 – 4 = 4 and 2 – 2 = 0, we obtain 40 as the first part of the answer. Then 3 – 7 forces borrowing. We merely reduce 40 by 1, obtaining 39, and conclude the calculation with 13 – 7 = 6 and 6 – 1 = 5 to obtain the answer, 3 965. Subtracting 1 a single time is easier than repeatedly borrowing, as would be necessary in manual calculation.

Note that it would be easy here to save time by calculating 82 – 42 = 40 rather than working digit by digit. Even in the first example, a good calculator might start with 82 – 43 = 39.

Add up from subtrahend

When I was a boy, children were taught to make change by tendering one coin or banknote at a time and reciting the running total up to the amount paid.[27] If the customer paid \$6.33 with a \$10 bill, the clerk would hand over a penny while saying «six thirty-four», then another penny while saying «thirty-five», and so on: «forty, fifty, seventy-five, seven dollars, eight, nine, ten». The customer could easily follow this transaction and see that the correct change from \$10 had been given, even though perhaps neither the clerk nor the customer calculated the amount of change explicitly.

Nowadays this simple skill has been largely lost, and clerks in stores are often quite unable to make change if the cash register fails....

This technique supports an additive alternative to subtraction: one adds numbers to the subtrahend until the minuend is reached.

In 2024, how old is a person born in 1976? The first thing to notice is that there is no need to perform a four-digit subtraction, since the answer will obviously have only two digits. We could subtract $124 - 76$, but it might be easier to start with 1976 and add 50 to get 2026, then subtract 2 to get 2024. The answer, then, is $50 - 2 = 48$ years.

Looking ahead, one could see that a number added to the final 6 to yield 4 would cause carrying, since $6 > 4$; that would justify adding only 40, to get 2016, and then 8, to get 2024, for an answer of $40 + 8 = 48$.

Another way: $1976 + 24 = 2000$, and $2000 + 24 = 2024$, so the difference is $24 + 24 = 48$.

If you struggle with subtraction, you may wish to try adding to the subtrahend instead. I find that it is not ideal for mental arithmetic, but you may like it.

---

27 The monetary examples here and in most other parts of this book presume US or Canadian dollars. Without wishing to seem parochial, I had to choose a currency. The examples may not be quite natural to you if your currency uses coins of 20 but not 25 units. Even in Canada, where the 1¢ coin was abolished years ago, the example in this section seems dated.

The complement

To subtract 999 999 from 5 000 000, you would not reach for paper and a pencil; you would instead subtract 1 000 000 and, since you had subtracted 1 too many, add 1 back:

$$\begin{aligned} 5\,000\,000 - 999\,999 &= 5\,000\,000 - (1\,000\,000 - 1) \\ &= 5\,000\,000 - 1\,000\,000 + 1 \\ &= 4\,000\,001. \end{aligned}$$

Many people, however, would manually calculate 500 − 87 even though they could proceed more simply:

$$500 - 87 = 500 - 100 + 13 = 413.$$

You may have noticed, almost without thinking, that 87 was 13 short of 100. If not, you need not worry: all will be explained in this section.

The *complement* of a number, when added to the number, brings the total to the next power of 10. Since 87 + 13 = 100, we say that 13 is the complement (properly the hundred's complement) of 87, and for that matter that 87 is the complement of 13. Similarly, the complement of 463 is 537, because 463 + 537 = 1 000 (or, equivalently, 1 000 − 463 = 537). Usually in mental arithmetic we compute complements from 100 or 1 000, but other complements are possible.

The complement is easy to calculate from left to right: simply subtract every digit in sequence from 9, except the last digit, which is subtracted from 10. (Trailing zeroes, however, are left unchanged; the last non-zero digit is subtracted from 10.) Since no digit subtracted from 9 or 10 can cause borrowing, one can breeze through the calculation of any complement. The complement of 77 is [9 − 7, 10 − 7] = 23; the complement of 618 is [9 − 6, 9 − 1, 10 − 8] = 382. Just remember to subtract the last digit from 10 even though each other digit is subtracted from 9.

Instead of subtracting a number directly, we can subtract the next power of 10 and add the complement. Thus 500 − 87 = 500 − 100 + 13 = 413, and 9 142 − 863 = 9 142 − 1 000 + 137 = 8 279.

Very often we use a partial complement by adding only up to the next *multiple* of a power of 10: thus 803 − 169 = 803 − 200 + 31 = 603 + 31 = 634.

Complements are most useful for avoiding borrowing; direct subtraction may be better when borrowing will not occur. For instance, 893 − 104 could be calculated as 893 − 200 + 96, but that would be harder than computing [8 − 1, 93 − 4] = 789 directly. For this reason, we tend to use the complement when the final digits of the subtrahend are higher than the corresponding digits of the minuend. In this instance, 93 is greater than 04; complements are therefore disadvantageous. In 803 − 169, by contrast, 03 is less than 69, and the complement will prevent two rounds of borrowing.

Practise complements until you can recognise all of the two-digit ones without thinking. You may already possess this skill.

Negative differences

If a is less than b, then $a - b$ is negative. It is most easily calculated as $-(b - a)$.

For example, 261 − 428 is negative because 261 < 428. It is the negative of 428 − 261 = 167, so −167.

Subtraction with large numbers

Four-digit subtraction is usually best done in pairs of digits. For example, 5 586 − 3 712 = [55 − 37, 86 − 12] = [18, 74] = 1 874.

There are two good ways to address borrowing that occurs across the pairs (at the third digit). First, when we notice that borrowing will be needed, we can perform it immediately. In

5 506 − 3 712, for instance, we may calculate 55 − 37 = 18 only to notice that 06 − 12 requires borrowing; we then reduce 18 to 17 and calculate instead 106 − 12 = 94, thereby obtaining 1 794 as the answer.

Second, we can accept negative numbers in our interim results:

$$\begin{aligned} 5\,506 - 3\,712 &= [55 - 37, 06 - 12] \\ &= [18, -6] \\ &= [17, 100 - 6] \\ &= 1\,794. \end{aligned}$$

In other words, «eighteen hundred negative six» is seventeen hundred ninety-four. This technique amounts to deferring our borrowing until the end of the subtraction, when we need to express the answer in standard numeric form. Just as we carry 100 leftward from a three-digit group too big for its two-digit space, we borrow 100 rightward into a two-digit group that has run negative and is therefore too small.

You do not have to use negative values in this way, but you may benefit from understanding the option to do so. Although our orthodox representation of numbers excludes internal negative components, the practical mental arithmetician will take full advantage of what Mathematics affords, even though eventually any calculations will have to be expressed in a conventional form. One can rightly take pleasure in deploying the number system flexibly and cleverly for a practical gain. Indeed, playful experimentation of this sort may prove instructive and even help to advance mathematical knowledge.

Beyond four digits, it is usually expedient to group the digits into threes, according to the English practice of representing numbers by their thousands, millions, billions, and so on. In the example of 673 084 – 229 515, we start with the thousands: 673 – 229 = 444. If no borrowing occurs, we can immediately declare «444 thousand ...» as the first part of the answer. A glance at the next part of the calculation, 084 – 515, shows that borrowing is necessary because 084 < 515; therefore, we instead declare «443 thousand ...» and borrow: 1 084 – 515 = 569. The answer is 443 569.

Chapter 5

Multiplication by a single digit

JUST AS WE LEARN to walk before we learn to run, so with mental arithmetic we solidify our skills at simple tasks before attempting complex ones. As a mental calculator, you will repeatedly multiply by a single digit. Developing the ability now will pay dividends later.

As with addition and subtraction, we always proceed from left to right. Calculating from right to left works well on paper, but it has no place in mental arithmetic.

Doubling

Multiplying by 2 is so important that we may as well give it a special name—doubling—and treat it separately from ordinary multiplication. You may already be good at doubling; if not, practise now until you can do it easily and naturally.

The only complication with doubling is the need to carry a 1 whenever the following digit is 5 or greater. Simply look ahead and add 1 when necessary. For example, $43 \cdot 2 = 86$, without carrying anything. In $46 \cdot 2 = 92$, doubling 4 yields 8, but the following 6 will cause carrying, so immediately increase 8 to 9; then doubling 6 yields 12, but the leading 1 has already been carried and can therefore be dropped, leaving 2 as the last digit.

In short, add 1 whenever the next digit is 5 or greater.

The following example illustrates the thoughts of a beginner with correct technique; the words in small capitals are the digits of the answer, the only ones that would be recited. A skilled calculator would not be so long-winded. In the beginning, however, speak aloud if you please; even perform your calculations on paper if you are not ready to attempt them mentally. Aim for accurate work and correct understanding. Mental proficiency and greater speed will both follow from mastery of the skills themselves.

$$239571408 \cdot 2 = 479142816$$

«Well, 2 doubled is FOUR; 3 doubled is 6, but the next digit is high, so make it SEVEN. Then 9 doubled is 18, but keep only the 8 because we have already carried 1; the next digit will cause carrying again, so we have not 8 but NINE. Next, 5 doubled is 10; make it 0, but ONE because of the following 7, which when doubled gives 14, so FOUR. Now, 1 doubled is TWO, and 4 doubled is EIGHT. Zero doubled is 0, but the following digit will make it ONE. Finally, 8 doubled, considering just the last digit, gives SIX.»

At the end, it would have been easier to double 08 as a group, getting 16 (read out as «ONE SIX» if the answer is given digit by digit). Likewise, one might have noticed that $14 \cdot 2 = 28$ and treated 14, too, as a group. For the moment, double each digit in succession until you have enough skill to find and use shortcuts.

With enough experience, you will find that you have learnt most or all of the doubles of two-digit numbers without trying. Then you will be able to double pairs of digits, rather than proceeding one digit at a time. This approach is especially good for pairs below 50, since there is nothing to carry:

$$\begin{aligned} 4183429576 \cdot 2 &= 4\ 18\ 34\ 29\ 5\ 76 \cdot 2 \\ &= 8\ 36\ 68\ 59\ 1\ 52. \end{aligned}$$

The first three groups were doubled directly; for 29 and 5, it was necessary to look ahead and add 1, as usual.

Halving

Similarly, division by 2 is so common and important that we shall call it halving. When halved, each even digit leaves no remainder; each odd digit leaves a remainder of 1 that will be carried as 10 added to the next digit. Halving simply reverses the procedure of doubling.

Try the previous example in reverse:

$$479142816 \cdot \frac{1}{2} = 239571408.$$

A skilled, if verbose, mental calculator might think in the following way:

«Half of 4 is TWO. Half of 7? Make it half of 6—THREE—and add 10 to the next digit, getting 19. Half of that is NINE, and the next number becomes 11; half of that is FIVE, and next is 14. Half of 14 is SEVEN; half of 2, ONE; half of 8, FOUR. Then there is a 1; half of it is ZERO, and the next digit becomes 16, which, when halved, yields EIGHT.»

Again, the end could have been simplified: $16 \cdot \frac{1}{2} = 08$. (Do not forget the zero.) Even $142816 \cdot \frac{1}{2}$ could have been done in pairs: $14\ 28\ 16 \cdot \frac{1}{2} = 07\ 14\ 08$. Attempt some of your halving in groups of digits once you can reliably and easily halve any number one digit at a time.

You may wish to practise halving digit by digit at first, but you should progress soon enough to partitioning numbers into two-digit groups that you can halve at once. The best groups are even numbers. If you do have an odd number, carry 50 to the next two-digit group.

The example above can be computed more efficiently:

$$479142816 \cdot \frac{1}{2} = 4\ 79\ 14\ 28\ 16 \cdot \frac{1}{2}$$
$$= 2\ 39\ 57\ 14\ 08.$$

Here 79 is odd, so we halved it (obtaining 39) and carried 50 forward to obtain $50 + 14 \cdot \frac{1}{2} = 50 + 7 = 57$. Note that 16 halved is 08, not 8.

Terminal zeroes

When one or both of the factors end in zeroes, simply remove all of those zeroes, multiply with the remaining digits, and append the zeroes to the answer. Thus $23 \cdot 20$ is $23 \cdot 2 = 46$ with a 0 appended, or 460; and $23\,000 \cdot 200$ is $23 \cdot 2$ with five zeroes appended, or 4 600 000.

Multiplying by 5

Because $5 = \frac{10}{2}$, multiplication by 5 is equivalent to multiplication by 10 with division by 2. In other words, advance the decimal point (or put a 0 onto the end if the number is an integer) and halve!

For instance, $8\,549 \cdot 5 = 8\,54\,90 \div 2 = 4\,27\,45 = 42\,745$. To multiply $26.781 \cdot 5$, we obtain $267.81 \div 2$, which can be computed as $(26\,78\,1) \div 2 = 13\,39\,05$. Noting the decimal point, we give the answer as 133.905.

Multiplying by 9

Because $9 = 10 - 1$, multiplying by 9 can be reduced to subtraction: $384 \cdot 9 = 384(10-1) = 384 \cdot 10 - 384 \cdot 1 = 3\,840 - 384 = [38 - 3, 40 - 84] = [38 - 4, 140 - 84] = 3\,456$.

Often it is easier to multiply by 9 directly, in the manner described in the next section. Try subtraction, however; you may prefer it, at least for certain classes of problems.

Multiplying by other digits

The procedure for the remaining digits resembles that of doubling, with two important differences. First, numbers greater than 1 may be carried. Indeed, one may have to carry a number as high as the digit minus 1. (This is why doubling—multiplication by 2—never requires the carrying of anything but 1.)

Second, carrying may cascade leftward. Consider the following example, calculated first as if on paper, then mentally:

$$378 \cdot 9 = 3\,402$$

Multiplying by hand using the standard algorithm from the first chapter, we would proceed from right to left: $8 \cdot 9 = 72$, put down 2, carry 7; $7 \cdot 9 + 7 = 70$, put down 0, carry 7; $3 \cdot 9 + 7 = 34$, which we write down to complete the answer.

Mentally, however, we always proceed from left to right. We start with $3 \cdot 9 = 27$; then $7 \cdot 9 = 63$, and we have $[27, 63] = [27 + 6, 3] = [33, 3] = 333$. Carrying thus affects *both* of the first two digits. Continuing, we get $8 \cdot 9 = 72$; now we have $[333, 72] = [333 + 7, 2] = 3\,402$. Again, carrying affects digits two positions back.

For this reason, we cannot always recite the digits as we calculate them. We *can* do so when doubling or halving, and sometimes when multiplying by another digit; but we should not be surprised if carrying occasionally changes a couple of digits that were previously calculated.

A good mental technique for multiplying two digits by one is to multiply the digits separately and combine the products without explicitly carrying anything. This is straightforward when there is nothing to carry: $32 \cdot 3 = [3 \cdot 3, 2 \cdot 3] = [9, 6] = 96$.

When the second product is 10 or more, carrying will be necessary; but it can managed by adding the tens' digit of the second product to the ones' digit of the first. Think as follows:

$$56 \cdot 3 = [15, 18] = [1, (5 + 1), 8] = 168$$
$$43 \cdot 8 = [32, 24] = [3, (2 + 2), 4] = 344$$
$$76 \cdot 9 = [63, 54] = [6, (3 + 5), 4] = 684$$
$$24 \cdot 4 = [8, 16] = [(8 + 1), 6] = 96$$

It may help you to think of the digits as being aligned in this way, using the example of $56 \cdot 3 = [15, 18] = 168$:

$$\begin{array}{r} 1\,5 \\ 1\,8 \\ \hline 1\,6\,8 \end{array}$$

With practice, you will think of the pair of digits in the middle and sum them. Try listing some 2×1 problems on paper and solving them by sight, writing the answer down (so that you can check it later) only when you have finished calculating it.

Again, sometimes it is necessary to carry a digit to the left after combining the middle digits:

$$47 \cdot 9 = [36, 63] = [3, (6 + 6), 3] = [3, 12, 3] = [4, 2, 3]$$
$$= 423$$
$$68 \cdot 8 = [48, 64] = [4, (8 + 6), 4] = [4, 14, 4] = 544$$
$$77 \cdot 7 = [49, 49] = [4, (9 + 4), 9] = [4, 13, 9] = 539$$
$$68 \cdot 3 = [18, 24] = [1, (8 + 2), 4] = [1, 10, 4] = 204$$

Rather than carrying twice, as in $47 \cdot 9$ above, you can merge any carried digit from the second figure into the whole of the first figure: $47 \cdot 9 = [36, 63] = [(36 + 6), 3] = [42, 3] = 423$. This strategy is no more difficult but does save a step.

Although this approach may seem awkward at first, you will quickly grow accustomed to it and see its advantages. Practise exclusively with two-digit numbers until you can multiply them quickly and accurately, almost automatically, by any single digit.

Larger numbers are multiplied similarly: the last digit of the partial answer is merged with the tens' digit (if any) of the following product. For example, to compute $427 \cdot 6$, we begin with $42 \cdot 6 = [24, 12] = 252$ (eventually you will streamline the process in this way); then $[252, 7 \cdot 6] = [252, 42] = 2\,562$. In the alternative, we can do all of the multiplication at once: $[24, 12, 42] = [25, 6, 2]$. This method of multiplication can be extended to any number of digits.

Once you can multiply two digits by one, try three digits, then four, then five. Much mental arithmetic involving large numbers depends on this skill; you should cultivate it before proceeding. Multiplying larger numbers will also increase your ability to carry several digits in your head—another skill essential to mental arithmetic.

Chapter 6

Squares of two or three digits

ALTHOUGH it can be treated as mere multiplication, the computation of squares enables us to exploit structure for speed and efficiency. Squaring can also simplify complex calculations such as the multiplication of large numbers and the extraction of square roots. The ability to square numbers quickly in one's head is both practically useful and satisfying.

The general method for squares

It is a fact of algebra that $(a+b)(a-b) = a^2 - b^2$:

$$\begin{aligned}(a+b)(a-b) &= a(a-b) + b(a-b)\\ &= a^2 - \cancel{ab} + \cancel{ab} - b^2\\ &= a^2 - b^2.\end{aligned}$$

By adding b^2 to each side of the equation, we obtain this expression for a^2:

$$a^2 = (a+b)(a-b) + b^2.$$

We simplify the work by choosing b so that either $a+b$ or $a-b$ ends in 0. A few examples will show just how easy this is.

In the following example, the nearest round number to 28 is 30; we therefore select b as the difference, namely 2, for a much simpler problem of multiplication:

$$\begin{aligned} 28^2 &= (28 + 2)(28 - 2) + 2^2 \\ &= 30 \cdot 26 + 4 \\ &= 780 + 4 \\ &= 784. \end{aligned}$$

In the next example, the nearest round number is lower than the number being squared. We can still let b be positive, since a negative value would not affect either $(a + b)(a - b)$ or b^2.

$$\begin{aligned} 33^2 &= (33 + 3)(33 - 3) + 3^2 \\ &= 36 \cdot 30 + 9 \\ &= 1\,080 + 9 \\ &= 1\,089. \end{aligned}$$

The nearest round number is never more than 5 away, which happens when the number ends in 5. An even easier technique explained below is used for that case, so b in practice need never be greater than 4.[28] There is nothing to carry when b is 1, 2, or 3, but 1 must be carried when b is 4:

$$\begin{aligned} 64^2 &= (64 + 4)(64 - 4) + 4^2 \\ &= 68 \cdot 60 + 16 \\ &= 4\,080 + 16 \\ &= 4\,096. \end{aligned}$$

This general method gives the square of any integer. Below are better methods that work for numbers with certain characteristics.

28 One may *choose*, however, to use a larger value of b. This would not ordinarily be advantageous for two-digit numbers, but it frequently is for larger numbers, as will be seen.

The square of a number that ends in zeroes

Square any number ending in one or more zeroes by omitting the zeroes, squaring what remains, and appending twice the number of zeroes. Thus 70^2 is $7^2 = 49$ with two zeroes appended, or 4 900; and 500^2 is 25 with four zeroes, or 250 000.

The square of a number that ends in 1

For any numbers a and b, we have $(a+b)^2 = a^2 + 2ab + b^2$:

$$\begin{aligned}(a+b)^2 &= (a+b)(a+b)\\ &= a(a+b) + b(a+b)\\ &= a^2 + ab + ab + b^2\\ &= a^2 + 2ab + b^2.\end{aligned}$$

This fact accounts for many of the special cases given below.

An integer ending in 1 can be written as $10a + 1$. Applying the equation above, we have

$$\begin{aligned}(10a+1)^2 &= (10a)^2 + 2(10a)(1) + 1^2\\ &= 100a^2 + 10(2a) + 1.\end{aligned}$$

Therefore, to square a two-digit number ending in 1, we drop the 1, obtaining a; then we merge $\left[a^2, 2a, 1\right]$, the components being single digits. Sometimes this is easy: $31^2 = \left[3^2, 2 \cdot 3, 1\right] = [9, 6, 1] = 961$. Often, however, carrying is required: $81^2 = [64, 16, 1] = [65, 6, 1] = 6\,561$.

The technique works for longer numbers as well: $131^2 = (13^2)100 + (2 \cdot 13)10 + 1 = [169, 26, 1] = [171, 6, 1] = 17\,161$.

The square of a number that ends in 9

It is also true that

$$(a-b)^2 = a^2 - 2ab + b^2.$$

To develop a method for squaring a number that ends in 9, we represent the number as $10a + 9$ and note that $10a + 9 = 10a + 10 - 1 = 10(a + 1) - 1$. Now we can apply the equation above:

$$\begin{aligned}\left(10(a+1)-1\right)^2 &= \left(10(a+1)\right)^2 - 2\left(10(a+1)\right)(1) + 1^2\\ &= 100(a+1)^2 - 10(2a) + 1.\end{aligned}$$

For example, since $79 = 80 - 1$, we obtain $79^2 = \left[8^2, -2\cdot 8, 1\right] = [64, -16, 1] = [62, 20 - 16, 1] = 6\,241$. Try a longer number: $249^2 = \left[25^2, -2 \cdot 25, 1\right] = [625, -50, 1] = [620, 0, 1] = 62\,001$.

The negative number in the middle complicates the calculation. In full, the calculations are $79^2 = 80^2 - 2 \cdot 80 + 1 = 6\,400 - 160 + 1 = 6\,241$ and $249^2 = 250^2 - 2 \cdot 250 + 1 = 62\,500 - 500 + 1 = 62\,001$.

You may find the general method easier for numbers that end in 9.

The square of a number that ends in 5

Numbers that end in 5 are especially easy to square, thanks to the fact that $(a + b)^2 = a^2 + 2ab + b^2$:

$$\begin{aligned}(10a+5)^2 &= (10a)^2 + 2(10a)(5) + 5^2\\ &= 100a^2 + (2\cdot 5\cdot 10)a + 25\\ &= 100a^2 + 100a + 25\\ &= 100a(a+1) + 25.\end{aligned}$$

This equation shows us how to square a number ending in 5: drop the final 5, multiply the remaining number by itself plus 1, and append 25. (The factor of 100 has the effect of making space for the two digits 25.)

Thus for 35^2, we drop the 5 to get 3; then we multiply 3 by itself plus 1, which is 4, to get 12; finally, appending 25 yields

1 225, the correct answer. Similarly, 85^2 is $8 \cdot 9 = 72$ followed by 25, or 7 225. For 395^2, calculate $39 \cdot 40 = 1\,560$; the answer, then, is 156 025.

The square of a number near 50

Numbers near 50 are also easy to square. To see why, calculate the square of $50 + a$:

$$\begin{aligned}(50 + a)^2 &= 50^2 + 2(50)(a) + a^2 \\ &= 2\,500 + 100a + a^2 \\ &= 100(25 + a) + a^2.\end{aligned}$$

For a value of a no greater than 9, add a to 25 and append a^2 as a two-digit number. (The factor of 100 gives space for a two-digit value of a^2, with nothing to carry.) Thus 53^2 is $25 + 3 = 28$ followed by $3^2 = 09$ (using two digits), or 2 809; and 59^2 is $25 + 9 = 34$ followed by $9^2 = 81$, or 3 481.

The equation above also holds for negative values of a. Thus $46 = 50 + (-4)$. Squaring it yields $25 + (-4) = 21$ followed by $(-4)^2 = 16$, or 2 116. (The negative sign matters in the first part of the calculation but not in the second, since the square of a negative number is positive. In other words, the square at the end will always be added, never subtracted.)

If $|a| \geq 10$, the square of a will have at least three digits, so carrying will be necessary: $62^2 = (50+12)^2$; we get $25+12 = 37$ followed by $12^2 = 144$, and the 1 must be carried, so $62^2 = 3\,844$. (In our simplified notation, $[37, 144] = [38, 44] = 3\,844$.) Likewise, $39^2 = (50-11)^2$; this is $25-11 = 14$ followed by $11^2 = 121$, and carrying gives 1 521 (or compute $[14, 121] = [15, 21]$).

For a number in the 30s, 40s, 50s, 60s, or 70s, add the last digit to 5, 15, 25, 35, or 45, respectively, and then add the square of the difference from 50: $38^2 = [5 + 8, 12^2] = [13, 144] = 1\,444$; $43^2 = [15 + 3, 49] = 1\,849$; $56^2 = [25 + 6, 6^2] = 3\,136$;

$67^2 = [35+7, 17^2] = [42, 289] = 4\,489$; $72^2 = [45+2, 22^2] = [47, 484] = 5\,184$.

The square of a number near 100

Similarly, to square $100 + a$, add $2a$ to 100 and append a^2 as two digits:

$$\begin{aligned}(100 + a)^2 &= 100^2 + 2(100)(a) + a^2 \\ &= 10\,000 + 200a + a^2 \\ &= 100(100 + 2a) + a^2.\end{aligned}$$

Thus 107^2 is $100+2\cdot 7 = 114$ followed by $7^2 = 49$, or 11 449, and 96^2 is $100 - 2 \cdot 4 = 92$ followed by $(-4)^2 = 4^2 = 16$, which is 9 216.

Outside the range 91–109, the method still works, but it entails carrying digits. For example, $88^2 = (100 - 12)^2$; we get 76 followed by 144, which, after carrying, is 7 744. Likewise, 118^2 gives 136 followed by 324, or 13 924.

For a number below 100, there is no need to compute $100 + 2a$ directly; instead, simply double the number and keep only the last two digits. In the example of 88^2, calculating $88 \cdot 2 = 176 \rightarrow 76$ is easier than calculating $100 - 2 \cdot 12$.

The square of a number near 75

Until you have learnt the squares up to 100 (from experience, not rote memorisation), you may like to calculate squares in the vicinity of 75 as follows:

$$\begin{aligned}(75 + a)^2 &= 75^2 + 2(75)(a) + a^2 \\ &= 5\,625 + 150a + a^2.\end{aligned}$$

Note that $150a$ is $100a$ plus half of $100a$.

Thus $79^2 = 5\,625 + 150 \cdot 4 + 4^2 = [56, 25] + [6, 16] = [62, 41] = 6\,241$, and $68^2 = (75 - 7)^2 = 5\,625 - 150 \cdot 7 + 7^2 = [56, 25] - [10, 50] + 49 = [56 - 10, 25 - 1] = 4\,624$.

The square of a three-digit number

The general technique for three-digit numbers resembles the one for two-digit numbers, except that we find the nearest multiple of 100. The difference from that multiple may be as great as 50, so knowing the squares up to 50 will prove beneficial.[29] This simple example involves no carrying:

$$\begin{aligned} 407^2 &= (407 - 7)(407 + 7) + 7^2 \\ &= 400 \cdot 414 + 7^2 \\ &= 165\,600 + 49 \\ &= 165\,649. \end{aligned}$$

The following example entails the slight additional complication of carrying:

$$\begin{aligned} 834^2 &= 800 \cdot 868 + 34^2 \\ &= 694\,400 + 1\,156 \\ &= 695\,556. \end{aligned}$$

For simplicity's sake, calculate $8 \cdot 868 = 6\,944$ and $34^2 = 1\,156$, then merge these numbers in two-digit chunks: $[6\,944, 1\,156] = [(6\,944 + 11), 56] = 695\,556$, or even just $[69, (44 + 11), 56] = [69, 55, 56]$.

---

29 In practice, the difference will not exceed 49, because a difference of 50 would imply that the last two digits of the number were 50, in which case we would calculate the square of a *two*-digit number and append two zeroes: $350^2 = 35^2 \cdot 10^2 = 1225 \cdot 100 = 122\,500$.

When the last two digits of the number exceed 50, the calculation proceeds similarly:

$$\begin{aligned}683^2 &= (683+17)(683-17)+17^2\\ &= 700\cdot 666+17^2\\ &= 466\,200+289\\ &= 466\,489.\end{aligned}$$

As in the previous section, the easiest way to calculate the difference is to double the number being squared and keep only the last two digits. Indeed, we need only double the last two digits of the number. Thus, instead of calculating $683 - 17 = 666$, we should have calculated $683 \cdot 2 \longrightarrow 666$ (because $83 \cdot 2 = 166 \longrightarrow 66$).

For most numbers, one can begin to announce the answer well before finishing the calculation. In the previous example, we might have said «Four hundred...» near the start, to impress others with our speed, and followed up with «... sixty-six thousand four hundred eighty-nine». In the following rare example, however, the first digit of the answer turns out at the last step to be higher than might have been expected:

$$\begin{aligned}837^2 &= 800\cdot 874+37^2\\ &= (6\,400+560+32)100+1\,369\\ &= 699\,200+1\,369\\ &= 700\,569.\end{aligned}$$

The square of a number near 500

The technique for squaring numbers near 50 can be extended to $(500+a)^2$: add a to 250 to obtain the thousands, then add a^2. If a^2 has no more than three digits, it can simply be appended as a three-digit number.

You can show off by giving the square of any number between 501 and 509 in a flash: simply say «Two hundred fifty», the last digit, «thousand», and the square of the last digit. For numbers from 491 to 499 inclusive, just start with «Two hundred forty» instead and end with the square of 10 minus the last digit. Thus $506^2 = 256\,036$ and $492^2 = 242\,064$.

Since a^2 can have three digits, $|a|$ can be as great as 31 without carrying, because $31^2 = 961$ is the last square with three digits. Greater values of $|a|$ will require carrying:

$$\begin{aligned} 542^2 &= (250 + 42)1\,000 + 42^2 \\ &= 292\,000 + 1\,764 \\ &= 293\,764. \end{aligned}$$

Similarly,

$$\begin{aligned} 463^2 &= (500 - 37)^2 \\ &= (250 - 37)1\,000 + 37^2 \\ &= 213\,000 + 1\,369 \\ &= 214\,369. \end{aligned}$$

To simplify the calculation even more, do not bother to add a to 250; for a number in the range 501–549, simply start the answer with 2, add 5 to the second digit, and copy the third digit. Numbers in the range 451–499 are handled similarly, except that 5 is subtracted, not added. Thus for 527^2, since $2 + 5 = 7$, we get 277 thousands plus $27^2 = 729$, or 277 729; and for 476 we get $226\,000 + 24^2 = 226\,576$. Even carrying does not greatly complicate the work: $545^2 = 295\,000 + 2\,025 = 297\,025$; $451^2 = 201\,000 + 2\,401 = 203\,401$.

The square of a number near 1 000

Squaring a number in the high 900s is easy indeed:

$$984^2 = 968 \cdot 1\,000 + 16^2$$

$$= 968\,000 + 256$$
$$= 968\,256.$$

As usual, compute not $984 - 16$ but $984 \cdot 2 \rightarrow 968$ (appending to the initial 9 the last two digits of $84 \cdot 2$). Call that number out, then say «thousand» and the square of the difference from 1 000. That is all!

A number less than 969 will have a difference whose square contains more than three digits and therefore requires carrying:

$$\begin{aligned} 951^2 &= 902 \cdot 1\,000 + 49^2 \\ &= 902\,000 + 2\,401 \\ &= 904\,401. \end{aligned}$$

Although ordinarily we use the nearest multiple of 100, the opportunity to use 1 000 is so attractive that we apply it to all numbers in the 900s—at least if we know the squares up to 100:

$$\begin{aligned} 912^2 &= (912 - 88)(1\,000) + 88^2 \\ &= 824\,000 + 7\,744 \\ &= 831\,744. \end{aligned}$$

The technique of doubling still works, but the first digit of any number in the 900s below 949 will be 8, not 9. In the example, $912 \cdot 2 \rightarrow 824$.

A number slightly *greater* than 1 000 is even easier to square, because $(1\,000 + a)^2 = 1\,000\,000 + (2a)(1\,000) + a^2$. If a is at most 31, we can impress people by announcing the answer immediately as «One million», then $2a$, then «thousand», then a^2: $1\,024^2 = 1\,048\,576$, because $24 \cdot 2 = 48$ and $24^2 = 576$. (Note that $24 \cdot 2 = 48$ must be written as 048.)

If a is as high as 414, the answer will still start with «One million», which we can therefore state immediately for an impressive display. Try $1\,414^2$:

$$1\,414^2 = 1\,000\,000 + (414 \cdot 2)(1\,000) + 414^2$$

$$= 1\,000\,000 + 828\,000 + (171\,200 + 196)$$
$$= 1\,000\,000 + (828 + 171)1\,000 + (200 + 196)$$
$$= 1\,000\,000 + 999\,000 + 396$$
$$= 1\,999\,396.$$

If you struggle to carry all of the digits in your head, rest assured that a later chapter presents a good way to remember numbers and that other chapters on squaring and multiplication explain how to use it for practical calculation.

In spare moments, practise squaring the numbers up to 50 in your head. Gradually you will come to know them by heart. Consider progressing then to the numbers from 51 to 100 until you know all of them as well. Knowledge of the first 50 squares, and preferably the first 100, will facilitate a range of calculations.

Chapter 7

Multiplication of two-digit numbers

QUITE OFTEN the need to multiply two-digit numbers arises in everyday life. Most people, however, can perform the calculation only on paper, if at all. Using the various methods in this chapter, you will be able to perform it quickly in your head, to the amazement of others. You will also have a skill that is essential for much of the material in subsequent chapters. You may even find the ability enjoyable and fulfilling.

More than most types of mental calculation, 2 × 2 multiplication opens ample space to the calculator's creativity. This lengthy chapter teems with techniques that you can apply in your own style. You may have your favourites, but all can be deployed profitably in some classes of problems. Learn them all, and practise calculating the same product in multiple ways.

Direct multiplication by a two-digit number

If you happen to know the multiplication table for a two-digit number, or even just the products that you need, you can multiply directly by that number as you would multiply by a single digit. The only complication—and it is minor—is that you may have to carry 10 or more.

For instance, if you know the multiplication table to 12×12 (as you may well if, like me, you are of a certain age), you can multiply by 12 directly: $47 \cdot 12 = 40 \cdot 12 + 7 \cdot 12 = 480 + 84 = 564$, and $89 \cdot 12 = 960 + 108 = 1\,068$.

If you live in the US or Canada, you may have learnt the first 9 multiples of 25, without noticing, from handling quarters (25¢ coins): for example, 7 quarters make \$1.75, so $25 \cdot 7 = 175$. Multiplication by 25 is then easy: $38 \cdot 25 = 750 + 200 = 950$.

Computer programmers, at least those of the old school, may know the 16s table from using hexadecimal (base 16) numbers. They can multiply directly by 16: $94 \cdot 16 = 1440 + 64 = 1\,504$.

You need not know an entire column of the multiplication table in order to multiply directly; knowledge of the facts that you need suffices. For instance, if you play bridge or certain other games, you know that a standard deck of cards has 52 cards distributed into 4 suits of 13 cards each. This knowledge, with the observation that $13 \cdot 3 = 39$, enables you to compute $34 \cdot 13 = 390 + 52 = 442$.

Perhaps you know from experience that there are 168 hours in a week. Then $24 \cdot 7 = 168$, and you can easily get $17 \cdot 24 = 240 + 168 = 408$.

A few devotees of mental arithmetic have gone so far as to learn the multiplication table up to 99×99. They can multiply four digits by four about as easily as we can multiply two digits by two. I do not recommend that you memorise thousands of products for the sake of facilitating mental calculation, but there are people who have done it. With experience, you are likely to expand your knowledge of the multiplication table from frequent use of the same products. I have come to know it up to 16×16, and most of it up to 20×20 or so. I also know the multiples of 23 up to 9 from calculations with logarithms that involve multiplication by $2.3023 = 23 \cdot 0.1001$. Any of this knowledge, of course, may be applied to general calculation.

Rearranging factors

The commutative and associative properties of multiplication allow us to arrange the factors in a product however we wish. This option is especially convenient for multiplying a number ending in 5 by an even number. For example,

$$\begin{aligned} 35 \cdot 46 &= 35(2 \cdot 23) \\ &= (35 \cdot 2)23 \\ &= 70 \cdot 23 \\ &= 7 \cdot 23 \cdot 10 \\ &= 161 \cdot 10 \\ &= 1\,610. \end{aligned}$$

When one number has two small factors, it may be expedient to multiply the other number by those factors in succession. One order, however, may be more convenient than the other. Consider $67 \cdot 36$: although 67 is prime, 36 can be factored as $6 \cdot 6$ or $9 \cdot 4$ or $12 \cdot 3$ or $18 \cdot 2$. Only the first three of these are convenient for calculation. Which should we use, and in which order should we multiply? If we take $4 \cdot 9$, we obtain

$$\begin{aligned} 67 \cdot 36 &= 67 \cdot 4 \cdot 9 \\ &= 268 \cdot 9 \\ &= 2\,412. \end{aligned}$$

See what happens when we take $9 \cdot 4$ instead:

$$\begin{aligned} 67 \cdot 36 &= 67 \cdot 9 \cdot 4 \\ &= 603 \cdot 4 \\ &= 2\,412. \end{aligned}$$

The 0 in the middle simplifies the calculation because there is nothing to carry: $603 \cdot 4$ is just $6 \cdot 4 = 24$ followed by $3 \cdot 4 = 12$.

The other three options—$6 \cdot 6$, $12 \cdot 3$, $3 \cdot 12$—also generate a 0 as the second digit because each of them contains a factor of 3 and $67 \cdot 3 = 201$. I would have noticed that and elected to multiply $67 \cdot 3 \cdot 12$.[30]

Likewise, $43 \cdot 56$ can be calculated as

$$\begin{aligned} 43 \cdot 56 &= 43 \cdot 8 \cdot 7 \\ &= 344 \cdot 7 \\ &= 2\,408 \end{aligned}$$

or, if you know that $43 \cdot 7 = 301$, much more easily as

$$\begin{aligned} 43 \cdot 56 &= 43 \cdot 7 \cdot 8 \\ &= 301 \cdot 8 \\ &= 2\,408. \end{aligned}$$

Doubling 51, 52, 53, and 54 yields 102, 104, 106, and 108, respectively. Look for an opportunity to double one of these numbers during multiplication. For example, $53 \cdot 24 = 106 \cdot 12 = 1\,272$.

With experience, you will learn some useful products with a 0 in the middle, such as $17 \cdot 6 = 102$, $18 \cdot 6 = 108$, $35 \cdot 3 = 105$, $34 \cdot 6 = 204$, and $88 \cdot 8 = 704$. You should not attempt to memorise lists of such facts; you can learn them more easily, and more thoroughly, through practice.

Also very helpful for simplifying calculations are $7 \cdot 73 = 511$ and $9 \cdot 79 = 711$. For example, $56 \cdot 73 = 8 \cdot 511 = 4\,088$ and $63 \cdot 79 = 7 \cdot 711 = 4\,977$.

30 You may already know that $\frac{2}{3} = 0.\overline{6}$ and that $\frac{2}{3} \cdot 100 = 66.\overline{6} = 66\frac{2}{3}$. The latter fact implies that $67 \cdot 3$ is slightly greater than 200: it is greater by $\frac{1}{3} \cdot 3 = 1$.

Using a point of reference

Applying the distributive property again, we obtain the following expression for the product of two numbers for which a is taken as a point of reference:

$$(a+b)(a+c) = a(a+b+c) + bc.$$

Note that $a + b + c = (a + b) + c = (a + c) + b$. We can therefore add c to $a + b$ (or, equivalently, add b to $a + c$), multiply by a, and add bc. This technique works best when both b and c are small.

The algebraic description does not convey the great practical utility of the technique, an important tool for any mental calculator. Several examples should illustrate its value.

To compute $13 \cdot 18$, we can take 10 as a common point of reference, since both factors are close to 10. We are therefore multiplying $(10+3)$ by $(10+8)$. We add 8 to 13, or else 3 to 18, as we wish, and obtain 21. This we multiply by the point of reference, 10, to get 210. Finally, we add $3 \cdot 8 = 24$, for a total of 234.

Instead, we can choose 20 as the point of reference; then we have $(20-7)(20-2)$. Calculating either $13-2$ or $18-7$ yields 11, which, multiplied by 20 (the point of reference), yields 220. Adding $(-7)(-2) = (7)(2) = 14$ gives the answer, 234.

When both factors are either above or below the point of reference, bc is positive; when one factor is above and the other is below, bc is negative. Consider $27 \cdot 32$. With 30 as the point of reference, we have $(30-3)(30+2)$. Since $27+2 = 32-3 = 29$, we calculate $29 \cdot 30 = 870$ and add $(-3)(2) = -6$ to obtain 864. Do not bother with negative numbers; just remember that you subtract when the point of reference falls between the factors and add otherwise.

Numbers that are not especially close to the point of reference may still be suitable for this method. For example, $33 \cdot 46$, with 30 as the point of reference, gives $30 \cdot 49 + 3 \cdot 16 = 1\,470 + 48 = 1\,518$. Likewise, $27 \cdot 46$, also with 30 as the point of reference, gives $30 \cdot 43 - 3 \cdot 16 = 1\,290 - 48 = 1\,242$.

When the numbers are below 100, there is no need to subtract from one or the other; simply add the numbers and keep only the last two digits. For example, $92 \cdot 87$ can be computed as $79 \cdot 100 + 8 \cdot 13 = 8\,004$, because $92 - 13 = 87 - 8 = 79$; however, you may find it easier to add $92 + 87 = 179$ and keep only the 79.

Practise this method with numbers that are reasonably close to a point of reference. Once you know how to use it, you will use it often. It is most effective when at least one of the numbers is close to the point of reference.

Difference of squares

The difference of two squares, especially when one of them is small, affords a sneaky but very effective and often fast way to perform multiplication.

In the section on squares, we applied the fact that

$$a^2 = (a + b)(a - b) + b^2.$$

For multiplication, by subtracting b^2 from each side of the equation, we can present the same fact as a difference of squares:

$$(a + b)(a - b) = a^2 - b^2.$$

This difference of squares can be handy, especially to those who know their squares. The chief difficulty lies in recognising that a product is in the form $(a+b)(a-b)$. One example is $18 \cdot 16$: if we notice that $18 = 17+1$ and $16 = 17-1$, we can calculate the product as $17^2 - 1^2 = 289 - 1 = 288$. Knowledge of the squares proves helpful again.

If the factors differ by an even number (which occurs when a and b are both even or both odd), let b be half of that number and a be the smaller factor plus b. The product is then $a^2 - b^2$. To multiply $28 \cdot 34$, for instance, we note that the factors differ by 6, an even number; therefore, we can take b to be $6 \div 2 = 3$ and a to be $28 + 3 = 31$ (or, equivalently, $34 - 3 = 31$). The product is $31^2 - 3^2 = 961 - 9 = 952$.

Note that a is the average of the factors: $(28 + 34) \div 2 = 62 \div 2 = 31$.

This method works best when b is small. For large b, it may be necessary to subtract multiple digits, and even to compute b^2 itself. Nevertheless, the method can be practical for multiplying two three-digit numbers when b is not much greater than 100 (*i.e.*, when the numbers do not differ by much more than 200). Try it.

If one factor is even and the other is odd, the method will not work directly; however, it can be adapted by subtracting 1 from one factor and adding the other factor at the end of the calculation. For instance, $39 \cdot 46 = 38 \cdot 46 + 46 = (42^2 - 4^2) + 46 = 1\,748 + 46 = 1\,794$. We could instead have calculated it as $39 \cdot 45 + 39 = (42^2 - 3^2) + 39 = 1\,755 + 39 = 1\,794$.

When you have mastered the difference of squares, the day will come when you will announce a product in a flash only to be asked by amazed bystanders how you calculated it. You may then modestly state «36 times 42 is 39 squared minus 3 squared, so 1 521 minus 9, or 1 512. Easy.»

Using subtraction

Sometimes it is convenient to multiply by a higher number and subtract the excess:

$$\begin{aligned} 74 \cdot 19 &= 74(20 - 1) \\ &= 74 \cdot 20 - 74 \cdot 1 \end{aligned}$$

$$
\begin{aligned}
&= 1\,480 - 74 \\
&= 1\,406.
\end{aligned}
$$

This method works especially well for multiplying by a number close to 100:

$$
\begin{aligned}
68 \cdot 96 &= 68(100 - 4) \\
&= 68 \cdot 100 - 68 \cdot 4 \\
&= 6\,800 - 272 \\
&= 6\,528.
\end{aligned}
$$

Notice how easy it is to multiply by 99: simply reduce the two-digit number by 1 and append its complement.

$$
\begin{aligned}
99 \cdot 32 &= 100 \cdot 32 - 1 \cdot 32 \\
&= 3\,200 - 32 \\
&= 3\,168.
\end{aligned}
$$

Think of it as $[32, -32] = [31, 100 - 32] = 3\,168$.

Usually the higher number is a round one, as in the previous examples; on occasion, however, another number might be used. The following example depends on the clever observation that $4 \cdot 75 = \cancel{4} \cdot \frac{3}{\cancel{4}} \cdot 100 = 300$, combined with the fact that 48 is a multiple of 4.

$$
\begin{aligned}
48 \cdot 74 &= 48(75 - 1) \\
&= 12 \cdot 4 \cdot 75 - 48 \\
&= 12 \cdot 300 - 48 \\
&= 3\,600 - 48 \\
&= 3\,552.
\end{aligned}
$$

Subtraction works best when one number either begins or ends with 9, as in the first two examples. It may not be the best choice for problems like this one:

$$
46 \cdot 68 = 46(70 - 2)
$$

$$= 46 \cdot 70 - 46 \cdot 2$$
$$= 3\,220 - 92$$
$$= 3\,128.$$

Try using subtraction and see how you like it. I tend to avoid it, except when multiplying by a number in the high 90s or when no better way presents itself.

Multiplication by 25, 50, or 75

To multiply by $50 = 100 \cdot \frac{1}{2}$, append two zeroes and halve the result: $86 \cdot 50 = 8\,600 \div 2 = 4\,300$; $69 \cdot 50 = 6\,900 \div 2 = 3\,450$. If the other factor is odd, the product will end in 50.

To multiply by $25 = 100 \cdot \frac{1}{4}$, append two zeroes and divide by 4: $68 \cdot 25 = 6\,800 \div 4 = 1700$; $79 \cdot 25 = 7\,900 \div 4 = 1\,975$. The remainders 1, 2, and 3 become the endings 25, 50, and 75, respectively.

To multiply by $75 = 100 \cdot \frac{3}{4}$, append two zeroes, multiply by 3, and divide by 4. The multiplication and the division can be performed in either order. For example, $75 \cdot 32 = 3 \cdot 3\,200 \div 4 = 3 \cdot 800 = 2\,400$; $93 \cdot 75 = 3 \cdot 9\,300 \div 4 = 27\,900 \div 4 = 6\,975$. As with multiplication by 25, the remainders 1, 2, and 3 correspond to the endings 25, 50, and 75, respectively.

Multiplication by 11

To multiply a two-digit number by 11, place between the digits their sum: $34 \cdot 11 = [3, 3 + 4, 4] = 374$. Sometimes there is a 1 to carry: $78 \cdot 11 = [7, 7 + 8, 8] = [7, 15, 8] = 858$.

Indeed, to multiply a number of any length, place between the first and last digits the sums of each successive pair of digits.[31]

31 A simpler formulation—add each digit to its neighbour on the right—is possible if we think of the number as being bounded by zeroes, as it is. Thus $258 \cdot 11 = 0258.0 \cdot 11 = [0 + 2, 2 + 5, 5 + 8, 8 + 0] = [2, 7, 13, 8] = 2\,838$.

For inſtance, $258 \cdot 11 = [2, 2+5, 5+8, 8] = [2, 7, 13, 8] = 2\,838$.

Work through the following example—as always, from left to right. An explanation of the technique follows.

$$\begin{aligned} 3\,975\,146 \cdot 11 &= [3, 3+9, 9+7, 7+5, 5+1, 1+4, 4+6, 6] \\ &= [3, 12, 16, 12, 6, 5, 10, 6] \\ &= [4, 3, 7, 2, 6, 6, 0, 6] \\ &= 43\,726\,606. \end{aligned}$$

As with doubling, the greateſt digit carried is 1, so the answer can be read off easily from left to right, digit by digit, if one looks ahead to anticipate any carrying. In the example, think as follows, and say only what is in small capitals: «Start with 3, but the next sum is 12, so make it FOUR; from the 12, after carrying, we have 2, but next comes 16, so THREE; now 6, but next is 12, so SEVEN; now TWO, with nothing to carry from the SIX, with nothing to carry from the 5; but 5 is followed by 10, so SIX; and we end with ZERO SIX.»

With a bit more effort, you can give the answer in ſtandard numeric form by noting that the answer has 1 digit more than the number being multiplied (though usually 2 more, because of carrying, if the number ſtarts with 9). For the 7-digit number above, the answer will contain 8 digits and will therefore be in the tens of millions; consequently, one says not «four» but «forty», not «three» but «three million», and so on.

Any other number whose two digits are the same can be factored as 11 multiplied by that digit. Try this example:

$$\begin{aligned} 43 \cdot 88 &= 43 \cdot 8 \cdot 11 \\ &= 344 \cdot 11 \\ &= 3\,784. \end{aligned}$$

Multiplication by 15

One good way to multiply by 15 involves the fact that $15 = 30 \div 2$. We can therefore multiply by 3, append a zero, and halve. Indeed, halving and multiplication can be done in either order. For instance, $28 \cdot 15 = 280 \cdot 3 \div 2 = 840 \div 2 = 420$; but easier is $280 \div 2 \cdot 3 = 140 \cdot 3 = 420$. With an odd number, halving will make the last digit a 5: $47 \cdot 15 = 470 \cdot 3 \div 2 = 1\,410 \div 2 = 705$.

Another good way is to note that $15 = 10 + 10 \cdot \frac{1}{2}$. We can therefore multiply by 10 and add half of the result: $28 \cdot 15 = 280 + 140 = 420$; $47 \cdot 15 = 470 + 235 = 705$.

A long number can be efficiently multiplied by 15 if the digits conveniently form even numbers: we add half to each group in succession. By looking ahead, we may be able to anticipate any carrying and call the answer out from left to right.[32]

$$\begin{aligned} 268941664 \cdot 15 &= 26\,8\,94\,166\,40\left(1 + \tfrac{1}{2}\right) \\ &= [26 + 13, 8 + 4, 94 + 47, 166 + 83, 40 + 20] \\ &= [39 + 1, 2 + 1, 41, 249, 60] \\ &= [40, 3, 41, 249, 60] \\ &= 4\,034\,124\,960. \end{aligned}$$

Note that the number of digits per group varies. We keep as many digits as each group contains and carry the rest leftward. Thus we carried the excess in $8 + 4 = 12$ and $94 + 47 = 141$ but not in $166 + 83 = 249$, because 166 has as many digits as the sum.

Multiplication by 37

Multiplying by 37 is easy if the other number is a multiple of 3, because $37 \cdot 3 = 111$: $37 \cdot 24 = 37 \cdot 3 \cdot 8 = 111 \cdot 8 = 888$.

32 See Ron Doerfler, «Developing a "Number Sense" through Mental Multiplication» at 2–3.

If two digits are left after factoring, treat them in the same way as a repeated «digit» and carry the excess. In the following example, the 3 from 32 in the second and third positions must be carried leftward:

$$\begin{aligned} 37 \cdot 96 &= 37 \cdot 3 \cdot 32 \\ &= [32, 32, 32] \\ &= [32 + 3, 2 + 3, 2] \\ &= [35, 5, 2] \\ &= 3\,552. \end{aligned}$$

Multiplication by a number beginning with 5

A number in the 50s can be profitably broken into 50 plus the last digit, because multiplying by 50 is the same as multiplying by 100 and halving.

$$\begin{aligned} 87 \cdot 53 &= 87(50 + 3) \\ &= 87 \cdot 50 + 87 \cdot 3 \\ &= 8\,700 \cdot \tfrac{1}{2} + 87 \cdot 3 \\ &= 4\,350 + 87 \cdot 3 \\ &= 4\,350 + 261 \\ &= 4\,611. \end{aligned}$$

This technique works especially well when the other number is even, for the product by 50 is one or two digits followed by two zeroes.

$$\begin{aligned} 64 \cdot 57 &= 64(50 + 7) \\ &= 6\,400 \cdot \tfrac{1}{2} + 64 \cdot 7 \\ &= 3\,200 + 448 \\ &= 3\,648. \end{aligned}$$

When all else fails

Usually one of the methods above will work well, but the all-purpose method never fails. It consists of breaking one of the numbers into its tens and its units, then multiplying by each and computing the total. This method is an application of the distributive property; indeed, it is the approach used for manual calculation.

$$\begin{aligned} 82 \cdot 38 &= (80 + 2)38 \\ &= 80 \cdot 38 + 2 \cdot 38 \\ &= 3\,040 + 76 \\ &= 3\,116. \end{aligned}$$

We could instead have broken 38 into its components; then we would have had $82 \cdot 38 = 82(30 + 8) = 2\,460 + 656 = 3\,116$. The answer is the same, of course, but the calculation seems more difficult because multiplying by 8 complicated the addition.

It is therefore generally better to start with the number whose units' digit is smaller, especially if that digit is 1:

$$\begin{aligned} 74 \cdot 61 &= 74(60 + 1) \\ &= 74 \cdot 60 + 74 \cdot 1 \\ &= 4\,440 + 74 \\ &= 4\,514. \end{aligned}$$

When one number starts with 1, consider breaking the other number:

$$\begin{aligned} 76 \cdot 17 &= (70 + 6)17 \\ &= 70 \cdot 17 + 6 \cdot 17 \\ &= 1\,190 + 102 \\ &= 1\,292. \end{aligned}$$

Breaking 17 instead would render both multiplication and addition more difficult: $76(10 + 7) = 760 + 532 = 1\,292$.

If the last digits are the same, break the larger number:

$$\begin{aligned} 68 \cdot 38 &= 60 \cdot 38 + 8 \cdot 38 \\ &= 2\,280 + 304 \\ &= 2\,584. \end{aligned}$$

In this situation, however, the difference of squares is likely to work better: $(53+15)(53-15) = 2\,809 - 225 = 2\,584$.

Examples for study

The following examples are meant to be instructive. Where there is no clearly superior technique, two or three options are presented and evaluated. Work through these examples yourself so that you can solidify your computational skills and also develop a knack for deciding quickly how to approach a two-digit multiplication problem.

$$43 \cdot 57$$

The difference of squares is far and away the best method: $50^2 - 7^2 = 2\,500 - 49 = 2\,451$. Look for this opportunity when the factors are equidistant from a round number, such as 50.

If not for that superior method, we might break 57, both because the 5 allows for halving and because $43 \cdot 7 = 301$ is easy to add: $43(50+7) = 43 \cdot 50 + 43 \cdot 7 = 2\,150 + 301 = 2\,451$.

$$88 \cdot 37$$

Subtraction works best: $(100-12) \cdot 37 = 100 \cdot 37 - 12 \cdot 37 = 3\,700 - 4 \cdot 3 \cdot 37 = 3\,700 - 444 = 3\,256$. Note that subtraction unveils a chance to use $37 \cdot 3 = 111$ for effortless multiplication.

Factoring works well if we multiply first by 11: $88 \cdot 37 = 8 \cdot 11 \cdot 37 = 8 \cdot 407 = 3\,256$. Multiplying by 8 first would be less expedient: $37 \cdot 8 \cdot 11 = 296 \cdot 11 = [2, 11, 15, 6] = 3\,256$.

$45 \cdot 82$

Double and halve: $45 \cdot 82 = 45 \cdot 2 \cdot 41 = 90 \cdot 41 = 3\,690$. This is substantially the same as breaking the number: $45(80 + 2) = 45 \cdot 2 \cdot 40 + 45 \cdot 2 = 90 \cdot 40 + 90 = 3\,600 + 90 = 3\,690$.

Factoring allows the use of the difference of squares: $45 \cdot 82 = 45 \cdot 41 \cdot 2 = 2(43^2 - 2^2) = 2(1\,849 - 4) = 2 \cdot 1\,845 = 3\,690$. Clearly this strategy is much more laboured than doubling to eliminate a terminal 5, which is quite likely to prove superior.

$83 \cdot 49$

Do not be tempted to factor 49 as $7 \cdot 7$: $83 \cdot 7 \cdot 7 = 581 \cdot 7 = 4\,067$. Subtraction works much better: $83 \cdot 49 = 83(50 - 1) = 83 \cdot 50 - 83 = 4\,150 - 83 = [41 - 1, 50 + 17] = 4\,067$.

The difference of squares is awkward: $66^2 - 17^2 = 4\,356 - 289 = [43 - 3, 56 + 11] = 4\,067$. By the time one found the squares to use, one could have multiplied by $50 - 1$.

$92 \cdot 97$

Use 100 as a point of reference: $100(92-3)+8 \cdot 3 = 8\,900+24 = 8\,924$. No other method should be considered.

$23 \cdot 77$

Note that $77 = 100 - 23$; therefore, we have $23(100 - 23) = 2\,300 - 23^2 = 2\,300 - 529 = [23 - 6, 71] = 1\,771$.

Factoring also works well: $23 \cdot 77 = 23 \cdot 7 \cdot 11 = 161 \cdot 11 = [1, 7, 7, 1] = 1\,771$.

The difference of squares is another option: $50^2 - 27^2 = 2\,500 - 729 = [25 - 8, 71] = 1\,771$. Although usually attractive when the first number is a multiple of ten, since the subtraction becomes a matter of appending a two-digit complement, the difference of squares may be less desirable for more distant numbers such as those in this example.

$76 \cdot 19$

If you notice that $76 = 19 \cdot 4$, you can rearrange the factors: $76 \cdot 19 = 38^2 = 1\,444$. Knowing the squares up to 50, you can retrieve the answer from memory.

Otherwise, direct multiplication by 19 works well, for those who can do it: $76 \cdot 19 = 70 \cdot 19 + 6 \cdot 19 = 1\,330 + 114 = 1\,444$. Subtraction is another good technique: $76 \cdot 19 = 76(20 - 1) = 1\,520 - 76 = [15 - 1, 20 + 24] = 1\,444$.

$53 \cdot 99$

Subtraction is almost always the best way to multiply a two-digit number by 99: $53(100 - 1) = 5\,300 - 53 = 5\,247$. The rule is to subtract 1 and append the complement: $[53 - 1, 100 - 53] = 5\,247$. (This works because $53 \cdot 99 = [53, -53]$.)

Often a number beginning with 5 can be profitably broken: $53 \cdot 99 = 50 \cdot 99 + 3 \cdot 99 = 4\,950 + 297 = [49 + 3, 50 - 3] = 5\,247$. Although this particular calculation is not difficult, the first technique given is far superior.

$23 \cdot 84$

Short of multiplying by 23 directly ($23 \cdot 80 + 23 \cdot 4 = 1\,840 + 92 = 1\,932$), the best method depends on the observation that the numbers differ roughly by a factor of 4. Doubling the first and halving the second reveals an opportunity to use the difference of squares: $23 \cdot 84 = 23 \cdot 2 \cdot 42 = 46 \cdot 42 = 44^2 - 2^2 = 1\,936 - 4 = 1\,932$.

This worked because the larger number was about 4 times the smaller *and* was a multiple of 4, so that it could share a factor of 2 while retaining one for itself. After redistribution of the factor of 2, the numbers met in the middle, and both were even, so the difference of squares succeeded.

We can also subtract from 25: $(25-2)84 = 25 \cdot 84 - 2 \cdot 84 = 25 \cdot 4 \cdot 21 - 2 \cdot 84 = 2\,100 - 168 = 1\,932$. Numbers near 25 (which is $100 \div 4$) or 75 (which is $300 \div 4$) lend themselves to this method when the other factor is a multiple of 4, as in this example.

$67 \cdot 62$

Perhaps the best method is to use 60 as a point of reference: $67 \cdot 62 = 60 \cdot 69 + 7 \cdot 2 = 4\,140 + 14 = 4\,154$.

Almost as good is to break 62, taking advantage of $67 \cdot 6 = 402$: $67 \cdot 62 = 4\,020 + 67 \cdot 2 = 4\,020 + 134 = 4\,154$. Look for an opportunity to use $67 \cdot 3 = 201$ or its variations.

We can set up an opportunity for the difference of squares: $67 \cdot 62 = (66 + 1)62 = 66 \cdot 62 + 62 = 64^2 - 2^2 + 62 = 4\,096 - 4 + 62 = 4\,154$. This method, however, is more complex and burdensome than the other two.

$74 \cdot 85$

That 5 invites doubling, but unfortunately the result is still a two-digit number: $74 \cdot 85 = 37 \cdot 170$. Continue with an easy difference of squares: $(27^2 - 10^2)10 = (729 - 100)10 = 6\,290$. This method is substantially the same as factoring: $37 \cdot 2 \cdot 17 \cdot 5 = 37 \cdot 17 \cdot 10$.

If you like to use a point of reference, you may do so here: $80(74+5) - 5 \cdot 6 = 80 \cdot 79 - 30 = 6\,320 - 30 = 6\,290$.

$16 \cdot 48$

Since $48 = 16 \cdot 3$, we have $16 \cdot 48 = 16^2 \cdot 3 = 256 \cdot 3 = [75, 18] = 768$. In the alternative, $16(50-2) = 16 \cdot 50 - 16 \cdot 2 = 800 - 32 = 768$.

$46 \cdot 83$

Factoring seems like barren ground at first, since 83 is prime; but $46 = 2 \cdot 23$, so we have an easy difference of squares: $2(23 \cdot 83) = 2(53^2 - 30^2) = 2(2\,809 - 900) = 2(1\,909) = 3\,818$.

Else we would probably break 83 up: $46(80+3) = 3\,680 + 138 = 3\,818$.

We could also subtract from 50: $(50-4)83 = 50 \cdot 83 - 4 \cdot 83 = 4\,150 - 332 = 3\,818$.

$36 \cdot 84$

Best is either $84 \cdot 12 \cdot 3 = [96, 48] \cdot 3 = 1\,008 \cdot 3 = 3\,024$ or $84 \cdot 6 \cdot 6 = [48, 24] \cdot 6 = 504 \cdot 6 = 3\,024$. The internal zero reduces the rest of the calculation to child's play.

The number 36 can be factored in several ways: $2 \cdot 18$, $3 \cdot 12$, $4 \cdot 9$, $6 \cdot 6$. Some of them are much less convenient in this problem: $84 \cdot 9 \cdot 4 = 756 \cdot 4 = 3\,024$ is not so bad, since $75 \cdot 4 = 300$ comes into play (and should be multiplied as a whole, not digit by digit); but $84 \cdot 4 \cdot 9 = 336 \cdot 9 = [27, 27, 54] = [27, 32, 4] = 3\,024$ (or $336 \cdot 9 = 336(10-1) = 3\,360 - 336 = 3\,024$) is decidedly inferior. In the absence of a good reason to do otherwise, we tend to multiply first by the larger factor, as multiplication often generates an extra digit and subsequent multiplication by a smaller value is likely to be easier.

Direct use of the difference of squares happens to be quite efficient; the difficulty lies in noticing its availability: $60^2 - 24^2 = 3\,600 - 576 = 3\,024$.

Breaking 84 up and reüsing $36 \cdot 4 = 144$ proves to be inferior: $36 \cdot 84 = 36(80 + 4) = 36 \cdot 4 \cdot 20 + 36 \cdot 4 = 144 \cdot 20 + 144 = 2\,880 + 144 = [28 + 1, 80 + 44] = [29, 124] = 3\,024$.

Factoring allows the use of a point of reference: $36 \cdot 42 \cdot 2 = 2\big(40(36+2) - 4 \cdot 2\big) = 2(40 \cdot 38 - 8) = 2(1\,520 - 8) = 2 \cdot 1\,512 = 3\,024$. This method is much more trouble than factoring in the first example. In general, avoid extra steps unless they confer a decisive advantage.

$39 \cdot 18$

This problem offers a variety of good approaches. It can be factored into a 2×1 product: $39 \cdot 2 \cdot 9 = 78 \cdot 9 = 702$. It can be factored for an easy final doubling: $39 \cdot 9 \cdot 2 = 351 \cdot 2 = 702$.

Subtraction can be used in either direction: $(40 - 1)18 = 720 - 18 = 702$ or $39(20 - 2) = 780 - 78 = 702$. A golfer who knows that there are 72 holes in four 18-hole rounds might use that knowledge here.

Factoring of 39 has less to recommend it: $13 \cdot 3 \cdot 18 = 13 \cdot 54 = 650 + 52 = 702$. It might have been a good choice if not for the better one of factoring 18.

Direct multiplication by 18 is efficient: $(30 + 9)18 = 30 \cdot 18 + 9 \cdot 18 = 540 + 162 = 702$. Note that this can be done as $3 \cdot 18 = 54$ and $9 \cdot 18 = 3(3 \cdot 18) = 3 \cdot 54 = 162$, then combined as $[54 + 16, 2] = 702$.

$66 \cdot 78$

Did you rush to take advantage of the factor of 11 ($78 \cdot 6 \cdot 11 = 468 \cdot 11 = [4, 10, 14, 8] = 5\,148$)?

If so, you missed an opportunity to turn a factor of 33 or 66 into 99. After studying the chapter on divisibility, you will easily see that 78 is divisible by 3. We can therefore reorganise the factors in this clever way:

$$\begin{aligned}66 \cdot 78 &= 33 \cdot 2 \cdot 26 \cdot 3\\ &= (33 \cdot 3)(26 \cdot 2)\\ &= 99 \cdot 52\\ &= 5\,200 - 52\\ &= 5\,148.\end{aligned}$$

Again, $99 \cdot 52 = [52, -52] = [52 - 1, 100 - 52] = [51, 48] = 5\,148$.

The difference of squares is good: $72^2 - 6^2 = 5\,184 - 36 = 5\,148$. Subtraction is not bad: $66(80-2) = 5\,280-132 = 5\,148$.

Conversion to 99 is the superior method, but it is harder to arrange than the others. Experience will build your ability to notice opportunities like this one.

Putting it all together

In another book,[33] I mention seeing the first twenty minutes or so of a film entitled *Gifted*. A girl on her first day of school disrupted class out of boredom (a familiar experience). The teacher challenged her with $57 \cdot 135$. Just as the teacher tired of waiting, the girl gave the answer, which the teacher confirmed on a calculator.

Go ahead and try this 3×2 problem even though only 2×2 multiplication has been discussed in this chapter. I shall give

33 *Chronicles of Cruising: How not to visit Iceland and Norway* (Edmonton: Horne Books, 2023) at 214–15. The book is a lightly edited collection of the daily accounts that I wrote for the amusement of friends on my efforts to make the most of a dreadful thirty-three-night mass-market cruise on which I was unfortunate enough to find myself.

several methods in the order in which they occurred to me, but please stop reading and attempt the calculation yourself first.

Here are my thoughts:

1) Ending in 5 as it does, 135 just begs to be doubled. Unfortunately, since 57 is odd, I cannot factor 2 out for that purpose (though see below[34]). Instead, I note that 57 is $19 \cdot 3$, and $3 \cdot 135$ gives the convenient result 405 (calculate it as $[39, 15] = [40, 5]$). On account of the zero in the middle, I can finish the problem without carrying anything: $19 \cdot 405 = [19 \cdot 4, 19 \cdot 5] = [76, 95] = 7\,695$.

2) Subtraction is also expedient: $57 \cdot 135 = (60 - 3)135 = 60 \cdot 135 - 3 \cdot 135$. Since $60 = 20 \cdot 3$, we can calculate $3 \cdot 135 = 405$ and then $60 \cdot 135 = 20 \cdot 405 = 8\,100$ to get $8\,100 - 405 = 7\,695$.

3) Although I would not usually multiply by 57 directly, in this case it is not difficult to do so: $57 \cdot 135 = 57 \cdot 100 + 57 \cdot 30 + 57 \cdot 5 = 5\,700 + 1\,710 + 570 \div 2 = 7\,410 + 285 = 7\,695$.

4) If we really want to try something exotic, we can return to that original idea of factoring 2 out of 57 so that we can double 135 to get 270. Because $57 \div 2 = 28\frac{1}{2}$, we now must calculate $270 \cdot 28\frac{1}{2}$. We can obtain $27 \cdot 28$ easily as $27 \cdot 4 \cdot 7 = 108 \cdot 7 = 756$ or almost as easily as $27^2 + 27 = 729 + 27 = 756$; thus $270 \cdot 28 = 7\,560$. To this we add $270 \cdot \frac{1}{2} = 135$ to obtain $7\,695$.

5) Since the factors are of equal parity (both are odd), we can apply the difference of squares: $57 \cdot 135 = (96 - 39)(96 + 39) = 96^2 - 39^2 = 9\,216 - 1\,521$. The subtraction is best done in two parts: $[92 - 15, 16 - 21] = [77, -5] = [76, 100 - 5] = 7\,695$. A valid technique, but more complicated because of the four-digit subtraction, which I would prefer to avoid.

34 Yes, already I was thinking of doing it anyway!

Reviewing this list now, I believe that I found these methods in order of decreasing desirability.

Doubling to knock the 5 out was the best idea, but it did not work.[35] Next best was factoring 57, which worked very well indeed. (Recall that the problem was not of my choosing; it came from a film. Had factoring not worked so well, or at all, I would still have found at least one way to perform the multiplication.)

Subtraction might have been a borderline technique if not for the opportunity to obtain the minuend essentially by doubling the subtrahend.

Multiplying by 57 directly was practical because the digits of 135 were small and convenient. Otherwise I probably would have computed $57 \cdot 130 + 57 \cdot 5$, especially since I know the multiples of 13. I did not consider $57 \cdot 100 + 57 \cdot 35$, although it is not difficult: use $57 \cdot 35 = (46+11)(46-11)$ or $57 \cdot 35 = 57 \cdot 70 \cdot \frac{1}{2}$. Nor did I consider $50 \cdot 135 + 7 \cdot 135 = 6\,750 + 945 = 7\,695$; it works, of course, but it seems more tedious than either direct two-digit multiplication or a breaking of the three-digit number.

That original idea of doubling was so attractive that I tried it even though I had to halve an odd number, thereby obtaining a fraction. Really, I tried it just for fun; I would not ordinarily multiply in this way.

The difference of squares, a favourite technique of mine, is less attractive here because the factors are so far apart that the squares have to be calculated: $(135 - 57) \div 2 = 39$ and $57 + 39 = 96$ (or $(57 + 135) \div 2 = 96$ and $96 - 57 = 39$). Then their four-digit squares have to be held in mind for subtraction. Even

35 Reviewing the manuscript for this book, Ron Doerfler observed that one could double at the beginning and halve at the end: $57 \cdot 135 = 57 \cdot 135 \cdot 2 \cdot \frac{1}{2} = 57 \cdot 270 \cdot \frac{1}{2}$. The difference of squares gives $57 \cdot 27 = 42^2 - 15^2 = 1\,764 - 225 = 1\,539$, so $57 \cdot 270 = 15\,390$. Then $15\,390 \cdot \frac{1}{2} = 7\,695$.

though subtraction is relatively easy, the calculation is cumbersome by this method. Some of the time I would have used this technique anyway, just because it would leave me feeling clever for a moment. Psychological considerations inform our choice of methods: we are not just number-crunching automata, after all!

You may not share my preferences. Find the methods that work best for you and use them when they work well. You may be able to deploy efficiently a method that I consider inferior, or you may dislike one that I find advantageous. Knowledge, experience, and taste all inform this difference of opinion. I am somewhat averse to subtraction as a support for multiplication, and I tend to avoid it; but some mental calculators use it skilfully and often. Conversely, knowing all of the squares up to 100, I use the difference of squares frequently (when the part to be subtracted is small!); but people who are less familiar with the squares might consider it less beneficial. In the example of $57 \cdot 135$, I quickly thought of factoring 57 as $19 \cdot 3$ and then multiplying 135 by 3 to yield a number with a 0 in the middle; someone who did not know that $57 = 19 \cdot 3$ probably would not have attempted factoring.

You now have a wealth of techniques with which to multiply two-digit numbers. Practise multiplying the same numbers in different ways, both to check your work (redo any erroneous calculation!) and to become familiar with the various techniques. Through practice, you will acquire a good sense of the right technique to use, so that you will not waste time pondering your options. You may not always find the ideal technique, but you should always find a good one.

Having chosen a technique to use, stick with it. Do not abandon a calculation just because it occurs to you that a different approach would have worked better.[36] If you have time, compute the answer in both ways: the experience will improve your ability to select a good technique, often the best one, right away.

36 All right, I admit that I must sometimes resist the urge to start over when in the middle of a calculation I notice that a different approach would have been superior, even clever. With my experience in mental arithmetic, I usually choose a good technique, perhaps the best one; yet on occasion I become annoyed with myself for missing a real humdinger.

Chapter 8

Divisibility

FACTORING, simplification of fractions, and some techniques for calculation depend on determining whether one number is divisible by another. Often there are better ways to test for divisibility than performing the division itself. This chapter presents tests for divisibility by all integers up to 12 and also suggests practical ways to check for divisibility by larger ones.

Every number is divisible by 1. No number is divisible by 0, since division by 0 does not exist. Zero itself, however, is divisible by every number, since the quotient is 0, with no remainder.

Divisibility by 10

A number that is divisible by 10 ends in a zero. For the same reason, a number that is divisible by 100 ends in two zeroes, a number that is divisible by 1 000 ends in three zeroes, and so on.

Divisibility by 5

A number that is divisible by 5 ends in 5 or 0.

Divisibility by 25

A number that is divisible by 25 ends in 25, 50, 75, or 00. You may be familiar with this fact from using 25¢ coins.

Divisibility by 2

A number that is divisible by 2 is even: it ends in 2, 4, 6, 8, or 0.

Divisibility by 4

In a number that is divisible by 4, the last two digits are divisible by 4. For instance, 3 824 is divisible by 4, because 24 is.

Divisibility can be determined from a glance at the last two digits: if the first of them is even, the second must be 0, 4, or 8; if odd, the second must be 2 or 6. Thus 4 716, 5 592, and 9 168 are divisible by 4; but 7 318, 2 990, and 6 042 are not.

Divisibility by 8

In a number that is divisible by 8, the last three digits are divisible by 8. Thus 3 824 is divisible by 8, because 824 is; 1 126, however, is not divisible by 8.

Divisibility by 3

In a number that is divisible by 3, the sum of the digits is also divisible by 3. Thus 3 284 is not divisible by 3, because $3+2+8+4=17$ is not; but 3 285 is divisible by 3, because its digits add up to 18, a multiple of 3.

Every 0, 3, 6, or 9, and every combination of digits whose total is a multiple of 3, may be disregarded. If all of the digits are thus eliminated, what remains—nothing—is equal to 0, which is divisible by 3; therefore, the original number is also divisible by 3.

To check 3 284, we need not add all of the digits: we can omit the 3, and also the 2 and the 4 (because they add up to 6, a multiple of 3). Only 8 remains, and it is not divisible by 3, so 3 284 is not divisible by 3 either.

For 3 285, we eliminate the 3 immediately. We can also eliminate the 2, 8, and 5, since their total is 15, a multiple of 3; then nothing remains, so we know that the number is divisible by 3. It might be easier, however, to stop looking for combinations of digits to eliminate and proceed instead with adding the digits up: $2+8+5 = 15 = 5 \cdot 3$, which establishes the divisibility of 3 285.

Divisibility by 9

In a number that is divisible by 9, the sum of the digits is also divisible by 9. Thus 3 284, whose digits add up to 17, is not divisible by 9; but 3 285 is, because its digits add up to 18, a multiple of 9.

Every 0 or 9, and every combination of digits whose total is a multiple of 9, may be disregarded.

Divisibility by 6

A number that is divisible by 6 is also divisible by both 2 and 3, the factors of 6. It must therefore be even, and the total of its digits must be a multiple of 3.

Divisibility by 12

A number that is divisible by 12 is also divisible by both 3 and 4, the factors of 12. Therefore, the total of its digits must be a multiple of 3, and its last two digits must be a multiple of 4.

Divisibility by 7

Most of the tests for divisibility by 7 that have been devised are slower and more cumbersome than division by 7 itself. They are of interest in number theory but not in mental arithmetic.

Sometimes division may be the easiest way to check for divisibility. Is 98 divisible by 7? Check: 9 ÷ 7 = 1 remainder 2; carry the 2 to make 28—and do not bother with the quotient, because 28 *is* divisible by 7, which is what we wanted to know.

Another method consists of adding or subtracting mutliples of 7 until obtaining a number of known divisibility. Along the way, we can discard trailing zeroes and also any *initial* string of digits that is divisible by 7. This works because adding or subtracting a multiple of 7 does not affect disability; nor does dropping trailing zeroes. An initial string that is divisible by 7 is a multiple of 7 multiplied by some number of factors of 10 and added to the rest of the number, so dropping it will not divisibility.

It is best to add or subtract a number that will make the last digit into 0. Gradually the number of digits that must be tested is reduced.

For example, we wish to know whether 56 322 is divisible by 7. First, 56 322 = 56 000 + 322; both 56 and 56 000 = 56 · 1 000 are divisible by 7, so we drop the 56 000 and check 322 for divisibility. We can subtract 42 to get 280 or add 28 to get 350; in either event, we drop the terminal zero and notice that the remaining number is divisible by 7; so, therefore, is 56 322 itself.

Try 73 428. First, the 7 can be dropped. (The 42 and the 28 cannot be dropped, since they, though divisible by 7, do not occur at the beginning.) We are left with 3 428. The last two digits are divisible by 7; we subtract them, leaving 3 400 and dropping the trailing zeroes to get 34. This number is not divisible by 0; neither is 73 428.

Divisibility by 11

One test for divisibility by 11 involves calculating the sum of the digits in odd-numbered positions (first, third, fifth ...) and subtracting the sum of the digits in even-numbered positions (second, fourth ...). If the number is divisible by 11, this difference will likewise be divisible by 11. For the example of 73 612, we calculate $7 + 6 + 2 = 15$ and $3 + 1 = 4$; the difference is 11, which is divisible by 11, so 73 612 is divisible by 11 as well. For 149 809, we have $1 + 9 + 0 = 10$ and $4 + 8 + 9 = 21$; the difference, -11, is divisible by 11, and so is the number itself.

At least in mental calculation, it may be easier simply to divide by 11 and check the remainder.

Special test for 7, 11, and 13

Because $1\,001 = 7 \cdot 11 \cdot 13$, subtracting any multiple of it from a number of four to six digits allows us to check for divisibility by 7, 11, and 13 simultaneously: we simply break the number into two three-digit groups and subtract the smaller from the larger, thereby reducing the number that must be tested to three digits or fewer.

In $11\,037 = [11, 037]$, the first group is smaller, so we subtract $37 - 11 = 26$. This we can easily check with division. It is not divisible by 7 or 11, but it is divisible by 13 ($26 \div 13 = 2$).

Not every number is quite so easy. Try 648 263. Since $263 < 648$, we compute $648 - 263 = 385$. You may recognise that 385 is divisible by 11 because the middle digit is the sum of the outer digits, but if necessary you would divide and find that $385 \div 11 = 35$. Is 385 also divisible by 7? Division yields $385 \div 7 = 55$, so the answer is yes; but if you had already calculated $385 \div 11 = 35$, you could have seen at a glance that 35 was divisible by 7, and therefore so was the original number. For the same reason, the number is not divisible by 13; but if necessary you could avoid dividing $385 \div 13$ by noting that 390 is divisible by 13 ($13 \cdot 3 = 39$), so $390 - 5 = 385$ is not.

In the second example, we are really calculating $648\,263 - 263 \cdot 1\,001 = 648\,263 - 263\,263 = 385\,000$. We can drop the trailing zeroes because $385 \cdot 1\,000$ is divisible by any of our three numbers if and only if 385 is. Therefore, we simplify the calculation to 648 − 263.

Similar tests are not hard to derive. Since $999 = 3^3 \cdot 37$, we can check for divisibility by 27 and 37 by adding the three-digit thousands to the three lower digits (thus turning a multiple of 999 = 1 000 − 1 into a multiple of 1 000 without the three zeroes). For 282 717, add 282 + 717 = 999. Clearly 999 is divisible by itself, so it is divisible by its factors. Likewise, $10\,001 = 73 \cdot 137$, so apply the trick for 1 001, using groups of four digits instead of three, if you ever need to check for divisibility by either of those factors. Using $2\,001 = 3 \cdot 23 \cdot 29$, subtract twice the lower group of three digits from the upper group: for 40 107, we get $40 - 107 \cdot 2 = 40 - 214 = -174$. Disregard the negative sign, which does not affect divisibility. The sum of the digits of 174 is 12, which is divisible by 3; 174 is not divisible by 23, but $29 \cdot 6 = 174$, so 40 107 is divisible by 3 and 29 but not by 23.

Divisibility by other numbers

To test for divisibility by a composite number, test for divisibility by each of its factors. Is 3 048 divisible by 24? If so, it is divisible by both 3 and 8.

For a composite number ending in 5, it may be convenient to apply doubling. Is 275 divisible by 15? If so, then $275 \cdot 2 = 550$ is divisible by $15 \cdot 2 = 30$, and (dropping the factor of 10) 55 is divisible by 3. But that is not true, so 275 is not divisible by 15.

Tests for divisibility by other prime numbers have been devised through the use of modular arithmetic; however, they can be difficult to remember, and they are not always convenient for mental calculation. For example, one test for divisibility by 17 involves subtracting 5 times the last digit from the number formed by the remaining digits, repeating the process if necessary, and checking for divisibility.[37] Take 11 037: subtracting $5 \cdot 7 = 35$ from 1 103 yields 1 068. Repeat: $106 - 5 \cdot 8 = 66$. If you recognise that 66 is not divisible by 17, you may stop; otherwise, repeat the process once more: $6 - 5 \cdot 6 = -24$, which is not divisible by 17.

Unless you need these tests frequently, you will find it easier to ignore them and perform division directly by focusing on the remainders, not the quotients. Using 11 037 again, we may be able to «divide» by 17 without thinking of the quotients directly, still less remembering them: for 110, the nearest multiple is 102 (never mind that it is $17 \cdot 6$); the difference is 8; for 83, the nearest multiple is 68, leaving 15; finally, $157 = 170 - 13$ is not divisible by 17 (or we can say that the nearest multiple is 153), so we have proven that 11 037 is not divisible by 17. Of course, if you expect to need the quotient, you should go to the slight extra trouble of remembering its digits.

37 This works because 51 is a multiple of 17. As in the chapter on squaring two- and three-digit numbers, $10a + b$ is a number whose last digit is b and whose other digits together are a: for instance, with $a = 1103$ and $b = 7$, we have $10a + b = 11\,037$. Suppose that $a - 5b$ is divisible by 17. Then so is $10(a - 5b) = 10a - 50b$. Likewise, adding $51b = 17(3b)$ does not affect divisibility, so we add and get $10a - 50b + 51b = 10a + b$, which is our original number. For the same reason, if $a - 5b$ is not divisible by 17, neither is $10a + b$. This proves the validity of our test for divisibility by 17.

Chapter 9

Division

NOT MANY OF US recall lessons on long division with undiluted pleasure. The process of finding a digit of a quotient is prone to error, and the work written below the dividend sends a diagonal cascade of digits down the page. The task of carrying so many figures in one's head for mental calculation may seem daunting. Fortunately, although we must sometimes resort to long division in mental arithmetic, usually there are better ways. Creative application of the various techniques can simplify what would otherwise be arduous work.

Recall that the remainder from division can be represented as a remainder or as a fraction: 20 417 ÷ 7 is either 2 916 R 5 or $2\,916\frac{5}{7}$. In addition, the fraction can be given as a decimal: $2\,916.\overline{714\,285}$. This chapter uses remainders for simplicity's sake.

Preparation

Begin by ensuring that the divisor is an integer without trailing zeroes. If it is not, move the decimal point in both divisor and dividend the same number of places to the left or right. Convert 12.3 ÷ 0.045 to 12 300 ÷ 45 by moving the decimal point three places forward; convert 12.3 ÷ 4 500 to 0.123 ÷ 45 by moving it two places backward.

How many digits?

The standard algorithm for long division reliably places the decimal point correctly. For mental arithmetic, lacking any visual point of reference, we may be wise to ascertain the position of the decimal point in advance. Starting from the left, take the smallest block of digits of the dividend that is greater than or equal to the divisor; the first digit of the quotient has the place value of the last digit of that block. In 20 417 ÷ 7, for example, we start with 20, because 2 alone is too small («7 won't go into 2»); thus the quotient begins in the thousands' place, like the 0 of 20, and is a four-digit number. In 0.002 041 7 ÷ 7, the first digit of the quotient is in the fourth decimal place (the ten thousandths' place) and therefore is preceded by 0.000.

Knowing the position of the decimal point, we can begin to recite the quotient before we have finished computing it. Seeing that 20 417 ÷ 7 was a four-digit number beginning with 2, we could promptly call out «2 thousand ...» and proceed with the calculation.

Calculate just enough digits

Unless it happens to come out even, division usually generates a heap of digits—more than are needed for most purposes. End your calculation when you have the number of digits that you need. For the last digit, you must determine whether the digit following it will cause rounding.

Short division

When the divisor is a single digit, or any other number by which we can divide directly, long division can be reduced to *short division*. Since each remainder must be less than the divisor (else the partial quotient is too small), the remainder will be a single digit whenever the divisor is a single digit. We can therefore do the multiplication and subtraction of long division mentally; the simple remainder will be the first part of the next partial dividend, and the digit that would be brought down will be its last digit. A simple example, 20 417 ÷ 7, illustrates the method:

$$\begin{array}{r} 2\;9\;1\;6\;\text{R}\,5 \\ 7\,\overline{)\,20{}^{6}4{}^{1}1{}^{4}7} \end{array}$$

Because 7 will not divide 2, we take 20 as the partial dividend and calculate 20 ÷ 7 as 2 remainder 6. The 2, written above the 0 of the dividend, is the first digit of the quotient; the remainder of 6 is joined to the next digit, 4, of the dividend to yield the new partial dividend, 64. As a crutch, we can write a small figure 6 next to the 4, as in this example.

Then 64 ÷ 7 is 9 remainder 1. The remainder and the next digit of the dividend form 11, which divided by 7 yields 1 remainder 4. Finally, 47 ÷ 7 is 6 remainder 5, and the answer is 2 916 remainder 5. (We could have continued by writing a decimal point after 2 916 and calculating 50 ÷ 7, then as many subsequent decimal places as we wanted.)

Practise this method on paper until you can write the answer directly, without recording the partial remainders:

$$\begin{array}{r} 2916\;\text{R}\,5 \\ 7\,\overline{)\,20417} \end{array}$$

Then develop the ability to recite the answer digit by digit while looking at the written problem, without writing anything. Soon you will be able to perform short division in your head even if the problem is not written down. With practice, you will find that the remainder from one partial quotient leads you straight into the next.

Usually short division is used with a one-digit divisor, but a larger divisor works well if we know its multiples up to 9. For example, we can easily divide by 11 directly:

$$\begin{array}{r} 1\ 8\ 5\ 6\,\mathrm{R}\,1 \\ 11\,\overline{)\,20^{9}4^{6}1^{6}7} \end{array}$$

Eleven times any digit is simply that digit repeated. For this reason, we get $20 = 1 \cdot 11 + 9$, $94 = 8 \cdot 11 + 6$, $61 = 5 \cdot 11 + 6$, $67 = 6 \cdot 11 + 1$. Knowledge of the multiplication table up to 12 makes direct short division by 12 possible.

Short division is an important technique that you can use in everyday life as well as in complex mental arithmetic. Practise it to the point of proficiency.

Cancel common factors

If the dividend and the divisor have a common factor, that factor may be cancelled from both of them to yield smaller numbers that may give rise to a simpler calculation. Instead of grinding $8\,415 \div 72$ out by long division, notice that $72 = 8 \cdot 9$ and that $8\,415$ too is a multiple of 9 (cast out $8 + 1 = 9$ and $4 + 5 = 9$ to obtain 0, which is divisible by 9!). You can then simplify the work by dividing 9 out of both numbers: $8\,415 \div 72 = 935 \div 8 = 116\,\mathrm{R}\,7$.

Even numbers frequently call out for this sort of simplification. Rather than dividing immediately by 2, see whether you can divide by 4 or 8: $7\,736 \div 48 = (967 \cdot 8) \div (6 \cdot 8) = 967 \div 6 = 161\,\mathrm{R}\,1$.

Indeed, if the divisor is even, consider dividing by 2 even if the dividend is odd: $7\,735 \div 48 = 3\,867.5 \div 24$ may be easier than the original problem. If we divide out three more factors of 2 (*i.e.*, 8), we get $483.4375 \div 3 = 161.1458\overline{3}$. Is there a faster way to divide by 48?

Sometimes cancellation can be repeated. In 13 176 ÷ 288, we might begin by noticing that both numbers are divisible by 8; then we have (13 176 ÷ 8) ÷ (288 ÷ 8) = 1 647 ÷ 36. Both of the new numbers are divisible by 9, so we can further simplify the problem: (1 647 ÷ 9) ÷ (36 ÷ 9) = 183 ÷ 4 = 45 R 3.

Cancelling common factors tends to reduce the number of digits and simplify the divisor—sometimes dramatically, as in the preceding example. It is the same technique that we use when simplifying fractions, which, after all, are just a form of division.

Note that changing the divisor will also change the remainder (unless it is 0). This will not matter unless you need the remainder as such; if you are treating it as a reduced fraction or a decimal, or abandoning it altogether, you need not worry. If necessary, multiply the calculated remainder by the factors that you cancelled.

Multiplying by a common factor

Cancelling factors just because one can, however, is not good policy. It may make sense to multiply divisor and dividend by a factor that will result in a simpler (though greater) divisor.

When both numbers end in 5, it is usually best to double them and drop the resulting zeroes: 2 905 ÷ 45 = 5 810 ÷ 90 = 581 ÷ 9 = 64 R 5. Often the divisor becomes one digit shorter. When it does not, it still tends to be easier to use in calculation: 2 905÷65 = 581÷13 = 44 R 9, and the remainder from division by 65 is $9 \cdot 5 = 45$.

Indeed, if only the divisor ends in 5, usually doubling is still advisable: 2 906÷65 = 5 812÷130 = 581.2÷13. This calculation can be continued in decimal form (44.707...) or stopped at 44 remainder 9.2; since the remainder equals $\frac{9.2}{13} = \frac{9.2 \cdot 5}{13 \cdot 5} = \frac{45+1}{65}$, division by 65 yields a remainder of 46.

Similarly, multiply by 4 when each number (or the divisor alone) ends in either 25 or 75: 2 925 ÷ 75 = 11700 ÷ 300 =

$117 \div 3 = 39$.

Multiplying throughout by 3 turns the divisor 37 into 111, which makes for very convenient division. Although 67 is a difficult divisor, multiplying by 3 transforms it into the easy figure 201. Doubling 56 to 112 may be expedient if division by a factor, such as 7 or 8, does not work.

Since $1\,001 = 7 \cdot 11 \cdot 13 = 7 \cdot 143 = 11 \cdot 91 = 13 \cdot 77$, a clever calculator can convert any of three awkward divisors to convenient 1 001 by multiplying: $42\,328 \div 143 = 296\,296 \div 1\,001 = 296(1\,001) \div 1\,001 = 296$. Although the absence of a remainder made this example particularly easy, the following example is only slightly more difficult: $59\,686 \div 143 = 417\,802 \div 1\,001 = (417\,417+385) \div 1\,001 = 417\frac{385}{1\,001} = 417\frac{55}{143} = 417\text{ R }55$.

Similar useful facts include $201 = 3 \cdot 67$; $301 = 7 \cdot 43$; $501 = 3 \cdot 167$; $801 = 9 \cdot 89$; $901 = 17 \cdot 53$; $1\,101 = 3 \cdot 367$; $1\,401 = 3 \cdot 467$; $1\,501 = 19 \cdot 79$; and $2\,001 = 3 \cdot 23 \cdot 29$.

Likewise, $299 = 13 \cdot 23$, so $4\,326 \div 23 = 56\,238 \div 299$. First, we divide by 300 instead: $56\,238 \div 300 = 187\text{ R }138$. Therefore, $56\,238 = 187 \cdot 300 + 138 = 187 \cdot 299 + (187 + 138) = 187 \cdot 299 + 325$. This divided by 299 is $187 + 325 \div 299 = 187 + 1 + 26 \div 299 = 188\text{ R }26$. Direct division by 23 might have been easier in this short example, but multiplication by 13 might have proven advantageous in a more difficult problem, or in one for which the subsequent division by 300 offered a sufficiently good approximation.

Similar useful facts include $399 = 3 \cdot 7 \cdot 19$; $799 = 17 \cdot 47$; $899 = 29 \cdot 31$; $1\,199 = 11 \cdot 109$; and $3\,999 = 3 \cdot 31 \cdot 43$.

Use this approach with care and wisdom. Taken too far, it will increase the number of digits without reducing the complexity of the calculation.

Dividing by a number ending in 9

Long division by 79 seems uninviting:

```
        258
 79 ) 20417
      158
      ---
       461
       395
       ---
        667
        632
        ---
         35
```

We can, however, take advantage of the fact that $79 = 80 - 1$. In the long division above, we subtracted $79 \cdot 2 = 158$. This amounts to subtracting $(80 - 1) \cdot 2 = 160 - 2$, which is the same as subtracting 160 and adding 2. In other words, we could have divided by 80 (which is tantamount to dividing by 8 with the decimal point shifted) and added the digit from the quotient to the next digit to be brought down from the dividend.

These observations give us a simplified way to divide by a number ending in 9: add 1 to obtain a divisor ending in 0; shift the decimal points so as to eliminate trailing zeroes; add each new digit from the quotient after bringing a digit down from the dividend.[38]

We can therefore calculate $20\,417 \div 79$ as follows:

```
       258.
 8 ) 2041.7
      46
       66
        3 5
```

We have changed 79 to 80, then shifted the decimal points: $2041.7 \div 8$. First, $20 \div 8$ gives 2 remainder 4; we bring down 4

38 See also AC Aitken, «The Art of Mental Calculation: With demonstrations», at 300.

to obtain 44 and then add the 2 from the quotient to obtain 46 (written below the dividend for reference). Then 46 ÷ 8 gives 5 remainder 6; bringing 1 down yields 61, to which we add the 5 from the quotient to obtain 66. Finally, 66 ÷ 8 is 8 remainder 2; with the 7 from the dividend, this is 27, plus the 8 just computed yields a remainder of 35.

If there is nothing to bring down, the work is even easier: simply append the new digit from the quotient to each remainder. We can calculate 17 ÷ 29 manually as follows:

$$\begin{array}{r|l} & 0.5\ 8\ 6\ 2\ 0\ \ldots \\ \hline 3\,) & 1.7\,^{2}5\,^{1}8\,^{0}6\,^{0}2\ ^{2}0 \end{array}$$

Originally we write the dividend (after adjusting the decimal point) as 1.7. Each new digit computed is entered into the quotient and, as a reminder, into the dividend as well. In this way, we proceed as with short division, by combining the previous remainder (either kept in memory or, as here, written as a small figure) with the next digit of the dividend.

For 17 ÷ 3, or 5 remainder 2, we write 5 into the quotient and at the end of the dividend as well. Then we merge the remainder of 2 with the next digit, 5, to obtain 25; divided by 3, this gives 8 remainder 1. Then 18 ÷ 3 is 6 remainder 0; 06 ÷ 3 is 2 remainder 0; 02 ÷ 3 is 0 remainder 2; and we could continue with 20 ÷ 3, which is 6 remainder 2, and as many other digits as we wanted.

Practise this calculation manually until you have a knack for it. You will soon acquire the ability to perform it in your head.

Dividing by a number ending in 1

A similar process enables easy division by 1; the only differences are that the divisor is reduced, not increased, by 1 and that each new digit of the quotient is subtracted, not added. The need to subtract does create an additional complication: the difference might be negative. Since we do not want any negative numbers in our quotient, we reduce by 1 any partial quotient that comes out even (*i.e.*, with remainder 0) and use the divisor as the remainder.

For example, 242 105 ÷ 61 = 3 968 R 57. We proceed as if dividing 24 210.5 by 6. First, 24 ÷ 6 is 4 remainder 0. Because the quotient came out even, we reduce it by 1 to 3; the remainder is then 6. We bring down the 2 of the dividend to obtain 62, then subtract the 3 just computed as the partial quotient to obtain 59. Then 59 ÷ 6 is 9 remainder 5; bringing down the 1 and subtracting the new 9 of the quotient, we have 51 − 9 = 42. Dividing by 6 yields 7 remainder 0, which we again convert to 6 remainder 6. Now we have 60 − 6 = 54, which, divided by 6, is 9 remainder 0; we convert this to 8 remainder 6. We bring down the final 5 of the dividend to obtain 65 and subtract the 8 of the quotient to obtain a remainder of 57.

Occasionally the expedient of reducing a remainderless quotient by 1 turns out to be wrong. In 365 944 ÷ 91 = 4 021 R 33, we have 36 ÷ 9 = 4 remainder 0, which we change to 3 remainder 9. Then 95 − 3 = 92 exposes our error, because 92 ÷ 9 is 10, which is too high for a one-digit quotient. We therefore restore the original partial quotient 4 and the remainder 0 to obtain 5 − 4 = 1. Dividing this by 9 yields 0 remainder 1. We bring the 9 down and obtain 19 − 0 = 19, which, divided by 9, is 2 remainder 1. Then 14 − 2 = 12; we have 1 remainder 3, and for the final remainder we obtain 34 − 1 = 33.

Reducing a remainderless quotient, however, is right much more often than not. When it proves to be wrong, the calculation is easily corrected.

❧

If there are no digits to bring down (other than zeroes), we know that the division will not terminate. We can therefore make a point of running 1 short—except at the end, where we add the 1 back to the remainder (if we care about the remainder at that point). This is the simplified process: subtract 1 to obtain a divisor ending in 0; shift the decimal points so as to eliminate trailing zeroes; subtract 1 from the first partial dividend before dividing; add the 9-complement of each new digit from the quotient after bringing a digit down from the dividend; divide and continue. Only the first partial dividend is reduced by 1; the 9-complement (*i.e.*, 9 minus the digit) makes the appropriate adjustment to all subsequent partial dividends. At the end, we correct the remainder by adding 1.

Note that this simplified process eliminates the need to reduce a partial quotient that comes out even. Use the exact quotient; it will be correct.

Here is how a skilled mental arithmetician would calculate $8 \div 41 \approx 0.195\,12$:

«Change the dividend to 0.8 and the divisor to 4. The quotient is therefore ZERO POINT something. Start by subtracting 1 from 8 to get 7; then $7 \div 4$ is ONE remainder 3. Next is 3 followed by $9 - 1$, or 38. That divided by 4 is NINE remainder 2, so we get 2 followed by $9 - 9$, or 20. Then $20 \div 4$ is FIVE remainder 0, and we have 0 and $9 - 5$, which is 4; and $4 \div 4$ is ONE remainder 0. That gives 0 and $9 - 1 = 8$, so 8, which, divided by 4, is TWO remainder 0. Let's stop here. The remainder is 0 followed by $9 - 2$, so 7—but we have to add 1 at the end to make up for the 1 subtracted at the beginning, so the corrected remainder is 8.»

Dividing by a number ending in 3 or 7

Multiplication by 3 converts a number ending in 3 to one ending in 9, or a number ending in 7 to one ending in 1. The techniques in the two previous sections can then be applied.

For example, $\frac{8}{23} = \frac{8 \cdot 3}{23 \cdot 3} = \frac{24}{69}$. Calculate 2.4 divided by 7, but with each partial quotient used as the last digit of the new partial dividend. Think as follows; the words in small capitals form the answer: «Well, 24 divided by 7 is 3, but there is a decimal point, so make it ZERO POINT THREE. The remainder is 3; append the partial quotient to get 33. Then 33 divided by 7 is FOUR remainder 5, and the next partial dividend is 5 followed by 4, or 54; divided by 7, this yields SEVEN remainder 5. Next, 57 divided by 7 is EIGHT remainder 1; 18 divided by 7 is TWO remainder 4; 42 divided by 7 is SIX even; 6 divided by 7 is ZERO remainder 6....»

Approximate division

If three or four digits' accuracy will suffice, a quickly computed approximation may take the place of an exact but laborious calculation.

During division, sometimes the denominator (divisor) differs slightly from a convenient value. When the denominator is too low by the small percentage $x\%$, consider multiplying numerator and denominator by $1 + x\%$; when it is too high by $x\%$, multiply by $1 - x\%$. This approximately corrects the value in the denominator, since

$$\left(1 + \frac{x}{100}\right)\left(1 - \frac{x}{100}\right) = 1 - \frac{x^2}{10\,000}$$

(by the difference of squares), which is close to 1 because $\frac{x^2}{10\,000}$ is small enough to be ignored. (If x has the relatively large value of 3, the omitted term will be 0.0009, which is less than 0.001; therefore, the approximation should be good to three digits if the substituted denominator is no more than 3% higher or lower.)

Multiplication explains why this technique works, but the calculation is even simpler: add or subtract any small percentage in both numerator and denominator! After all, multiplying a number by $1 + x\%$ just adds $x\%$ to the number, and multiplying by $1 - x\%$ just subtracts $x\%$.

For example, $23\,456 \div 793$ looks difficult, but it can be estimated quickly to three or four digits' accuracy:

$$\begin{aligned}\frac{23\,456}{793} &= \frac{23\,456}{800-7}\\ &= \frac{23\,456}{800(1-7/800)}\\ &\approx \frac{23\,456}{800(1-1\%)}\\ &= \frac{23\,456(1+1\%)}{800(1-1\%)(1+1\%)}\\ &\approx \frac{23\,456(1+1\%)}{800}\\ &= 29.32(1+1\%)\\ &\approx 29.32+0.29\\ &= 29.61.\end{aligned}$$

The correct answer, to four digits, is 29.58. Had we used $1-\frac{7}{8}\%$ rather than the computationally easier but less accurate estimate $1-1\%$, we would have obtained not three but four correct digits.

When the divisor is slightly too high, we subtract a percentage from the numerator:

$$\begin{aligned}\frac{23\,456}{254} &= \frac{23\,456}{250+4}\\ &\approx \frac{23\,456}{250\left(1+1\frac{1}{2}\%\right)}\\ &\approx \frac{23\,456\left(1-1\frac{1}{2}\%\right)}{250}\\ &= \frac{93\,824\left(1-1\frac{1}{2}\%\right)}{1000}\\ &= 93.824\left(1-1\tfrac{1}{2}\%\right)\\ &\approx 93.824-1.407\end{aligned}$$

$$= 92.42.$$

To four digits, the answer is 92.35. Had we taken 4 to be 1.6% of 250 rather than 1.5%, we would have obtained 92.32, which is closer to the correct value.

The calculations are written out above with multiplication just to reënforce conceptual understanding. In practice, they are performed as follows (using the simplified notation $x \pm y\%$ to mean $x \pm x \cdot y\%$, which is x plus or minus $y\%$ of x):

$$\begin{aligned}\frac{23\,456}{793} &= \frac{23\,456}{800-7}\\ &\approx \frac{23\,456}{800(1-1\%)}\\ &\approx \frac{23\,456}{800} + 1\%\\ &= 29.32 + 1\%\\ &\approx 29.32 + 0.29\\ &= 29.61.\end{aligned}$$

The second example above, in simplified form:

$$\begin{aligned}\frac{23\,456}{254} &= \frac{23\,456}{250+4}\\ &\approx \frac{23\,456}{250\left(1+1\tfrac{1}{2}\%\right)}\\ &\approx \frac{23\,456}{250} - 1\tfrac{1}{2}\%\\ &= 93.824 - 1\tfrac{1}{2}\%\\ &\approx 93.824 - 1.407\\ &= 92.42.\end{aligned}$$

In the first example, we would notice that 793 was about 1% shy of 800; we would therefore add about 1% to the numerator and divide by 800, or divide first and then add. In the second example, we would notice that 254 was about $1\frac{1}{2}$% above 250; we would subtract about $1\frac{1}{2}$% from the numerator and divide by 250, or reverse the order of the operations.[39]

In day-to-day life, often a three- or four-digit estimate of a quotient is good enough. Many of the sophisticated calculations later in this book likewise need only that degree of accuracy, or less. Approximate division often turns a nasty job of long division into an impressively quick estimate, at the cost of a small loss of accuracy.

Long division streamlined

When all else fails, we may have to perform a calculation that is tantamount to mental long division. With its cavalcade of digits, long division can strain the memory. It is therefore best to reorganise the work for convenient mental arithmetic.

```
          267
764 ) 204175
      1528
      ----
       5137
       4584
       ----
        5535
        5348
        ----
         187
```

39 Reversing the order works here because both the dividend and the small adjustment are divided by the new divisor, so we can perform the addition or the subtraction first or second. Ordinarily, of course, we cannot exchange the order of division and addition or division and subtraction.

Rather than computing 204 175 ÷ 764 by conventional long division, we shall break the work into easier steps that yield the same result.

We start with four digits of the dividend: 2041 ÷ 764 gives the partial quotient 2. Rather than computing $764 \cdot 2 = 1\,528$ all at once, we subtract piecemeal. From the 20 of the dividend we subtract 2 times the first digit of the divisor, 7: $20 - 14 = 6$. To this 6 we append the next digit of the dividend, 4, to obtain 64, from which we subtract 2 times the next digit of the divisor, 6: $64 - 12 = 52$. We append 1 from the dividend and subtract $2 \cdot 4$: $521 - 8 = 513$. Note that now the remainder is that of subtracting $2041 - 1528 = 513$ in the tableau above.

Continuing, we bring down 7 to make 5137 and obtain 6 as the next digit of the quotient. Subtraction proceeds as before: $51 - 6 \cdot 7 = 51 - 42 = 9$; $93 - 36 = 57$; $577 - 24 = 553$. Bring down 5 to produce 5535. The next partial quotient is 7. For the remainder, $55 - 49 = 6$; $63 - 42 = 21$; $215 - 28 = 187$. We could continue by writing a decimal point, bringing down a 0 to obtain 1870, and so on. Instead, we stop at 267 R 187. (Incidentally, if we do not care about the remainder, we need not bother to compute it after finding the last digit of the quotient.)

Setting the work up in this way greatly reduces the number of digits that we must carry in our heads. We multiply only single digits and perform subtraction incrementally.

Try 92 235 ÷ 473. We see that 922 ÷ 473 is not quite 2, so the quotient begins with 1. Now we may wish to subtract $922 - 473$ directly to obtain 449; but we could subtract $9 - 4 = 5 \rightarrow 52 - 7 = 45 \rightarrow 452 - 3 = 449$. In either event, we bring down 3 to make 4493. The next partial quotient is 9. Subtract: $44 - 36 = 8 \rightarrow 89 - 63 = 26 \rightarrow 263 - 27 = 236$. Bringing down the 5 makes 2365. The next digit of the quotient is 5. Subtract: $23 - 20 = 3 \rightarrow 36 - 35 = 1 \rightarrow 15 - 15 = 0$. The quotient is 195 exactly.

Chapter 10

Fractions and decimals

KNOWING the decimal representations of common fractions is useful in its own right, but it can also enable you to simplify difficult calculations by substituting the form that is more convenient. This chapter presents techniques for converting between fractions and decimals, as well as applications of those techniques to division.

The most commonly used fractions

The decimal expansions of fractions are best learnt in groups. Easiest are the tenths, since $\frac{n}{10} = 0.n$:

$$\frac{1}{10} = 0.1; \quad \frac{2}{10} = 0.2; \quad \frac{3}{10} = 0.3\ldots.$$

These give us the fifths as well:

$$\begin{array}{ccc|ccc} \frac{1}{5} = \frac{2}{10} & & 0.2 & \frac{3}{5} = \frac{6}{10} & & 0.6 \\ \frac{2}{5} = \frac{4}{10} & & 0.4 & \frac{4}{5} = \frac{8}{10} & & 0.8 \end{array}$$

The eighths include the quarters and $\frac{1}{2}$:

$\frac{1}{8}$	0.125	$\frac{5}{8}$	0.625
$\frac{1}{4} = \frac{2}{8}$	0.25	$\frac{3}{4} = \frac{6}{8}$	0.75
$\frac{3}{8}$	0.375	$\frac{7}{8}$	0.875
$\frac{1}{2} = \frac{2}{4} = \frac{4}{8}$	0.5		

The ninths include the thirds. Like the tenths, they exhibit an obvious pattern: $\frac{n}{9} = 0.\overline{n}$.

$\frac{1}{9}$	$0.\overline{1}$	$\frac{5}{9}$	$0.\overline{5}$
$\frac{2}{9}$	$0.\overline{2}$	$\frac{2}{3} = \frac{6}{9}$	$0.\overline{6}$
$\frac{1}{3} = \frac{3}{9}$	$0.\overline{3}$	$\frac{7}{9}$	$0.\overline{7}$
$\frac{4}{9}$	$0.\overline{4}$	$\frac{8}{9}$	$0.\overline{8}$

Three of the sixths are fractions given above: $\frac{2}{6} = \frac{1}{3} = 0.\overline{3}$; $\frac{3}{6} = \frac{1}{2} = 0.5$; $\frac{4}{6} = \frac{2}{3} = 0.\overline{6}$. The two remaining values:

$$\frac{1}{6} = 0.1\overline{6}; \quad \frac{5}{6} = 0.8\overline{3}.$$

The elevenths also exhibit a pattern. They need not be learnt one by one: $\frac{1}{11} = 0.\overline{09}$, so $\frac{n}{11} = 0.\overline{(9 \cdot n)}$.

$\frac{1}{11}$	$0.\overline{09}$	$\frac{6}{11}$	$0.\overline{54}$
$\frac{2}{11}$	$0.\overline{18}$	$\frac{7}{11}$	$0.\overline{63}$
$\frac{3}{11}$	$0.\overline{27}$	$\frac{8}{11}$	$0.\overline{72}$
$\frac{4}{11}$	$0.\overline{36}$	$\frac{9}{11}$	$0.\overline{81}$
$\frac{5}{11}$	$0.\overline{45}$	$\frac{10}{11}$	$0.\overline{90}$

The sevenths may appear challenging, but they simply repeat the cycle 142857 starting from different digits:

$\frac{1}{7}$	$0.\overline{142857}$	$\frac{4}{7}$	$0.\overline{571428}$
$\frac{2}{7}$	$0.\overline{285714}$	$\frac{5}{7}$	$0.\overline{714285}$
$\frac{3}{7}$	$0.\overline{428571}$	$\frac{6}{7}$	$0.\overline{857142}$

These fractions are also useful:

$$\frac{1}{12} = 0.08\overline{3}$$

$$\frac{1}{15} = 0.0\overline{6}$$

$$\frac{1}{16} = 0.0625$$

$$\frac{1}{18} = 0.0\overline{5}.$$

The twelfths are easy. Knowing $\frac{1}{12} = 0.08\overline{3}$, one can get $\frac{5}{12} = 0.41\overline{6}$; then $\frac{7}{12} = 1 - \frac{5}{12} = 0.58\overline{3}$ and $\frac{11}{12} = 1 - \frac{1}{12} = 0.91\overline{6}$. The other twelfths are equivalent to simpler fractions whose values appear above.

Converting fractions to decimals

The techniques for division that are explained in the previous section are good for mental calculation of the decimal expansions of fractions, especially if they are carried to only three or four digits. When calculating the *exact* decimal expansion, one can benefit from the following knowledge.

All fractions have a repeating decimal pattern (the *repetend*), which may be $\overline{0}$ (*e.g.*, $\frac{1}{2} = 0.5\overline{0}$.).[40] If the denominator of the fraction in lowest terms is n, the *period*—the number of digits in the repetend—is at most $n - 1$: division can yield only n possible remainders (from 0 to $n - 1$) and 0 as a remainder cannot occur in any repetend other than $\overline{0}$ (since it ends the division); thus the first remainder will occur again, restarting the repetend, after $n - 1$ digits at most. In addition, $n - 1$ is always divisible by the period. Of the fractions listed above, $\frac{1}{4}$ ($0.25\overline{0}$), $\frac{5}{6}$ ($0.8\overline{3}$), and $\frac{4}{9}$ ($0.\overline{4}$) are among those with only one repeating digit; but $\frac{1}{7}$ has 6 ($0.\overline{142857}$), and $7 - 1 = 6$ is divisible by 6; $\frac{5}{11} = 0.\overline{45}$, with 2 digits, and $11 - 1 = 10$ is divisible by 2.

Except in special cases such as those with n divisible only by factors of 2, 3, or 5, there is no good way to determine the period *a priori*. The fraction $\frac{1}{97}$ has the maximum period of 96, whereas $\frac{1}{101}$ has a period of only 4.

A useful fact for a denominator that is not divisible by 2, 3, or 5 and whose repetend has an even period is that the digits in the second half of the repetend are those from the first half, each

40 As is discussed in the first chapter, a number followed by $\overline{9}$ is equivalent to the same number, increased by 1 in the last digit, followed by $\overline{0}$. Thus there are two ways to write any number with a «terminating» decimal. By convention, the one with $\overline{0}$ is almost always used, with the $\overline{0}$ omitted.

subtracted from 9. For example, the period of $\frac{1}{7}$ must be a factor of $7 - 1 = 6$, so the only possibilities are 2, 3, and 6 itself.[41] When we compute the decimals by dividing $1 \div 7$, we get 0.142 with a remainder of 6. There are more digits to come, because the remainder 6 has not appeared before—but we can stop dividing now, because we know that the period is not 2 or 3, so it must be 6. Since 6 is even and the denominator 7 is not divisible by 2, 3, or 5, the remaining digits are 1, 4, 2 each subtracted from 9, so the decimal expansion of $\frac{1}{7}$ is $0.\overline{142857}$.

When division produces a partial dividend that has occurred previously at or after the final non-zero digit in the numerator, there is no need to continue, because the decimals will repeat and therefore the repetend has been found. In the example of $\frac{1}{7}$, division begins with $10 \div 7$; after six digits, division yields a remainder of 1, which becomes a partial dividend of 10 when a 0 is brought down, so any further calculation would repeat the work already done, and the six digits that have been calculated are the repetend. For $\frac{1}{3}$, the initial dividend is 10; so is the next one, owing to a remainder of 1, and so the repetend is the single digit 3 that has already been obtained.

A fraction whose denominator has only factors of 2 or 5 has a «terminating» decimal—one whose repetend is $\overline{0}$ and therefore is ordinarily omitted. One factor of 2 combines with one factor of 5 to yield 10, which contributes a single zero to the beginning of the decimal expansion; the remaining 2s or 5s contribute one digit each to the end. Therefore, the decimal expansion will contain exactly as many digits as the number of factors of 2 or of 5, whichever is greater.

If the denominator contains a factor other than 2 or 5, the fraction has a non-zero repeating decimal.

41 The only denominators with a one-digit repetend are divisible by 2, 3, or 5; since 7 is divisible by none of these, we can dismiss 1 from consideration as a factor.

The short forms of division by a number ending in 1 or 9 are well adapted to calculating decimal expansions. Denominators ending in 3 or 7 can be multiplied by 3 for use with these techniques.

Converting decimals to fractions

A «terminating» decimal is already in fractional form, the denominator being 1 followed by a 0 for each decimal digit (*i.e.*, 10^n for n digits). Moreover, a «terminating» decimal is easily reduced to lowest terms, because the denominator, consisting of nothing but factors of 10, is divisible by 2 and 5 only. The numerator can quickly be tested for these factors. For example, $0.003\,125 = \frac{3\,125}{1\,000\,000}$; the numerator is divisible by 25 because it ends in 25, and indeed we find that it is 5^5; removing 5 factors of 5 from both numerator and denominator yields $\frac{1}{320}$, in lowest terms. Likewise, $0.012 = \frac{12}{1\,000}$; the numerator is $2^2 \cdot 3$, and removing 2 factors of 2 yields $\frac{3}{250}$, which is in simplest terms because, as we knew at the outset, the denominator is not divisible by 3.

Other decimals require more effort to convert to fractions in lowest terms. We exploit the fact that multiplication by some number of factors of 10 will place the repeating decimal such that it can be subtracted away. To convert $0.\overline{714\,285}$ to a reduced fraction, proceed as follows:

$$\begin{aligned} x &= 0.\overline{714\,285} \\ 1\,000\,000x &= 714\,285.\overline{714\,285} \\ 1\,000\,000x - x &= 714\,285 \\ 999\,999x &= 714\,285 \\ x &= \frac{714\,285}{999\,999} \end{aligned}$$

$$
\begin{aligned}
&= \frac{26\,455 \cdot \cancel{3^3}}{37\,037 \cdot \cancel{3^3}} \\
&= \frac{715 \cdot \cancel{37}}{1\,001 \cdot \cancel{37}} \\
&= \frac{65 \cdot \cancel{11}}{91 \cdot \cancel{11}} \\
&= \frac{5 \cdot \cancel{13}}{7 \cdot \cancel{13}} \\
&= \frac{5}{7}.
\end{aligned}
$$

Finding some of those common factors is admittedly not easy, unless one notices (remember the trick for multiplying by 37?) that $999 = 9 \cdot 111 = 27 \cdot 37$ and knows or discovers that $1001 = 7 \cdot 11 \cdot 13$. This example, however, is more complex than anything that would likely be undertaken through mental calculation; it serves to illustrate what is possible. A more typical example:

$$
\begin{aligned}
x &= 0.\overline{592} \\
1\,000x &= 592.\overline{592} \\
999x &= 592 \\
x &= \frac{592}{999} \\
&= \frac{16 \cdot \cancel{37}}{27 \cdot \cancel{37}} \\
&= \frac{16}{27}.
\end{aligned}
$$

Percentages

A percentage is just a fraction whose denominator is 100. Consequently, $0.136 = 13.6\%$ and $\frac{7}{25} = \frac{7 \cdot 4}{25 \cdot 4} = \frac{28}{100} = 28\%$.

A 20% discount on a \$60 purchase results in a savings of $\$60 \cdot 20\% = \$60 \cdot 0.20 = \$12$. The discounted price can be calculated directly: $60 \cdot 100\% - 60 \cdot 20\% = 60 \cdot (100 - 20)\% = 60 \cdot 80\% = 60 \cdot 0.80 = 48$, or \$48. Similarly, to add a 5% tax to \$24, we can add the tax of $24 \cdot 5\% = 1.20$ to 24, or we can calculate $24 \cdot (100 + 5)\% = 24 \cdot 105\% = 24 \cdot 1.05 = 24 + 1.20 = 25.20$; either way, the total cost is \$25.20.

Note that $x\%$ of y is $y\%$ of x, because both are $\frac{xy}{100}$. Thus we can change 36.5% of 20 to 20% of 36.5 if we find that easier.

In mental arithmetic, we can often exploit the percentage by which one number differs from a more convenient one. Estimating this percentage is less difficult than it may seem. For example, if we must calculate with 6.872, we may prefer to substitute 7 and make an adjustment for the difference. That difference is approximately -0.13 (using 6.87 instead of 6.872 because we seek only a rough approximation of the percentage), and $\frac{0.13}{7} \approx \frac{0.14}{7} = 0.02$, which is 2%. Using the simplified notation of this book, which is intended to foster conceptual facility with mental arithmetic, we can therefore write $6.872 \approx 7(1 - 2\%)$ or even just $6.872 \approx 7 - 2\%$ (which means 7 minus 2% *of 7*, not 7 minus the fixed amount 0.02). Similarly, 90.5 is about $\frac{1}{2}\%$ greater than 90, because $\frac{0.5}{90} = \frac{0.05}{9} \approx 0.005 = \frac{1}{2}\%$; thus $90.5 \approx 90 + \frac{1}{2}\%$.

Division by decimals

Often division by decimals affords a good opportunity to use the technique, presented in the previous chapter, of eliminating a factor of $1 \pm x$ from the divisor by multiplying dividend and divisor (or, equivalently, numerator and denominator) by $1 \mp x$.

In the following example, we would prefer to divide by 8. Since $7.83 = 8 - 0.17$ and $0.17 = \frac{17}{100} \approx \frac{16}{100} = 8 \cdot \frac{2}{100} = 8 \cdot 2\%$, we proceed with the following approximation:

$$\frac{49}{7.83} = \frac{49}{8 - 0.17}$$

$$\begin{aligned}
&\approx \frac{49}{8(1-2\%)}\\
&\approx \frac{49(1+2\%)}{8(1-2\%)(1+2\%)}\\
&\approx \frac{49}{8}(1+2\%)\\
&= 6.125(1+2\%)\\
&= 6.125 + 0.1225\\
&= 6.2475.
\end{aligned}$$

As is explained above, only the first three digits can be assumed to be accurate. We therefore round the answer to 6.25, which is correct to three digits.

More simply put, $49 \div (8 - 2\%) \approx (49 \div 8) + 2\%$. We correct the shortfall of 2% in the denominator by adding 2% to the numerator, or to the calculated quotient.

The technique works well when the denominator is close to a decimal that we recognise as a fraction. For example, to divide by 7.58, we could note that the figure is about 1% higher than 7.5, which is $10 \cdot \frac{3}{4} = \frac{30}{4}$; therefore, we could perform the division approximately by multiplying by $\frac{4}{30}$ (the reciprocal of $\frac{30}{4}$) and subtracting 1%.

Recognition of the decimal expansion of a fraction enables us to calculate the radius of the Earth quite easily.[42] The circumference is 40 000 km.[43] Since the circumference C of a circle

---

42 The Earth bulges slightly at the Equator, so it is not spherical. The deviation is so slight, however, that the Earth is commonly treated as a sphere for a broad range of scientific and technological purposes. The «radius» is calculated as the average of measurements taken at different points on the surface; the maximum difference is about $\frac{1}{3}$%.

43 The original definition of the metre gave the distance from the Equator to the North Pole along a meridian passing through Paris as 10 million metres. The distance all the way around the Earth is 4 times that figure, which is 40 million metres or 40 000 kilometres.

is 2π times the radius r, we find that

$$r = \frac{C}{2\pi} \approx \frac{40\,000}{2 \cdot 3.14} = \frac{40\,000}{6.28}.$$

If we recognise that $\frac{1}{16} = 0.0625$, we can see that the denominator is close to $\frac{100}{16} = 6.25$; it exceeds 6.25 by 0.03, which is approximately $\frac{1}{2}\%$ (since $\frac{3}{6.25} \approx \frac{3}{6} = \frac{1}{2}$). Thus the radius of the Earth is approximately $40\,000 \cdot \frac{16}{100} - \frac{1}{2}\%$. Otherwise, we still notice that the decimal in the denominator is close to 0.25, so we multiply the denominator by 4:

$$\begin{aligned} 6.28 \cdot 4 &= (6.25 + 0.03) \cdot 4 \\ &= (24 + 1) + 0.12 \\ &= 25.12. \end{aligned}$$

Because 25.12 is close to 25, we multiply by 4 to obtain 100.48, which is approximately $\frac{1}{2}\%$ higher than 100. Division is reduced to the trivial task of moving the decimal point back two places.

In either event, we obtain the same result:

$$\begin{aligned} \frac{40\,000}{6.28} &\approx \frac{40\,000 \cdot 16}{100 + \frac{1}{2}\%} \\ &\approx \frac{640\,000 - \frac{1}{2}\%}{100} \\ &= \frac{640\,000 - \frac{1}{2} \cdot 6\,400}{100} \\ &= 6\,400 - 32 \\ &= 6\,368. \end{aligned}$$

Because we used a three-digit approximation of π and also introduced a slight error when dividing, we can expect only the first three digits of our result to be accurate. We therefore round 6 368 to 6 370. Standard works of reference give the average radius of the Earth as 6 371 km, so we calculated it correctly to three digits.

Decimals in multiplication

In the previous example, the fact that $\frac{1}{16} = 0.0625$ proved useful for what would ordinarily have been a difficult problem of division. Knowledge of fractions and their corresponding decimals is also useful in multiplication. To multiply by 875, note that $0.875 = \frac{7}{8}$; therefore, $875 = \frac{7}{8} \cdot 1000$, and one can proceed easily by multiplying by 7, appending three zeroes, and dividing by 8. (If the multiplicand is divisible by 8, it is usually better to divide first by 8, then multiply by 7 and append the zeroes.)

The repeating decimals are less useful, but now and then one can find a clever application for them. Multiplying by 67 is difficult, since the digits are large and the number is prime. The following example, however, shows a practical use for the decimal expansion of $\frac{2}{3}$.

$$\begin{aligned}
348 \cdot 67 &= 348(66.\overline{6} + 0.\overline{3}) \\
&= 348\left(\frac{2}{3} \cdot 100 + \frac{1}{3}\right) \\
&= 116 \cdot 3\left(\frac{2}{3} \cdot 100 + \frac{1}{3}\right) \\
&= 116(2 \cdot 100 + 1) \\
&= 116 \cdot 200 + 116 \cdot 1 \\
&= 23\,200 + 116 \\
&= 23\,316.
\end{aligned}$$

This example reflects the fact, mentioned in the chapter on 2×2 multiplication, that $67 \cdot 3 = 201$. More simply put, $348 \cdot 67 = 116 \cdot 3 \cdot 67 = 116 \cdot 201$.

Repeating decimals can also be useful for estimates. To multiply *approximately* by 83, recall that $\frac{1}{12} = 0.08\overline{3}$; therefore, $83 \approx \frac{1}{12} \cdot 1000$, so simply append three zeroes and divide by 12 to two digits. For greater accuracy, add an estimate of $\frac{1}{3}$ times the multiplicand.

Mixed numbers and improper fractions

A *mixed number*, such as $3\frac{1}{2}$ or $8\frac{7}{15}$, combines an integer with a fraction.

Especially when multiplying or dividing, it is often best to convert any mixed number to an *improper fraction*—a fraction greater than or equal to 1. The numerator of the improper fraction is the denominator of the mixed number multiplied by the integral part, plus the numerator; the denominator is that of the mixed number. For instance, $2\frac{7}{8} = \frac{8 \cdot 2+7}{8} = \frac{23}{8}$.

To convert an improper fraction to a mixed number, simply divide: $\frac{23}{8} = 23 \div 8 = 2\frac{7}{8}$. It may be necessary to reduce the fractional part to lowest terms: $\frac{22}{8} = 2\frac{6}{8} = 2\frac{3}{4}$.

Chapter 11

The calendar

OUR AWKWARD CALENDAR saddles us with inconsistencies. The months, even the quarters, are of different lengths. The days of the week shift from year to year. Leap years offer an additional complication.

In the middle of the twentieth century, a reformer named Elisabeth Achelis led a valiant campaign to institute her World Calendar. Each quarter on her calendar has 91 days, therefore 13 weeks. The first month in each quarter has 31 days; the other two, 30 each. The 365th day was to be «Worldsday», intended as a holiday—New Year's Eve. Worldsday was an intercalary day, not assigned to any day of the week and also not found in any month. In leap years, an extra intercalary day was to be added after June. Every day occurred on a consistent day of the week, year after year. Each year, even each quarter, started on a Sunday.

The World Calendar achieved broad international support but was scuppered by the UN in 1955 thanks—predictably—to the US, which opposed it on the grounds that the intercalary days would disrupt the seven-day cycle of some religions' holy days. Various religious authorities supported the new calendar; some rigid believers, however, insisted on having their holy days occur *exactly* every seven days—and forcing the rest of us to maintain a broken calendar for their convenience. For this reason, the UN shelved the project, and to this day we have to wrangle with our clumsy, archaic calendar out of deference to real or imagined religious tradition.

Structure of the Gregorian calendar

Today we use the Gregorian calendar, which was introduced by the Roman Catholic Church in 1582 and gradually adopted all over the world over the next four centuries. The Gregorian calendar accounts for the mean duration of the solar year, about 365.242 20 days, by adding an occasional leap day to its 365-day plan. Every year divisible by 4 is a leap year, except those that are divisible by 100 but not by 400. Thus 1700, 1800, and 1900 were not leap years, but 2000 was one.

With this adjustment, the Gregorian calendar repeats itself every 400 years; therefore, any day is on the same day of the week as the same date 400 years ago. The average duration of the year over that 400-year interval is 365.242 5 days, which is slightly too high: the world will probably have to skip a leap day in 2000 years or so, perhaps in the year 4000.

The Gregorian calendar replaced the Julian calendar, which over a period exceeding a thousand years had accumulated ten extra days, such that it no longer corresponded to the seasons. Those ten days were simply dropped from the calendar: 4 October 1582 was followed by 15 October 1582. Britain did not adopt the calendar until 1752; it had to drop eleven days from the month of September.[44] The different dates of adoption resulted in anomalies: Cervantes and Shakespeare both died on 23 April 1616, but Cervantes died ten days earlier, because Spain was already on the Gregorian calendar but England still used the Julian; the October Revolution succeeded on 7 *November* 1917,

44 Some people alleged a conspiracy to shorten their lives. More reasonable was the concern about having to pay the whole month's rent for only 19 days.

which was 25 October under the Julian calendar (which the Bolsheviks promptly changed to the Gregorian).

The 365 days of the year form 52 seven-day weeks, plus 1 extra day (2 days in a leap year); thus the day of the week for a given date advances by 1 or 2 days from one year to the next. The months are likewise of irregular lengths: April, June, September, and November have 30 days each; the other months have 31, except February, with 28 (29 in a leap year).

Adding and subtracting days

In my work as a lawyer, it often happens that a document must be filed at the courthouse within 20 days of an event. I make a note of the deadline. Sometimes it is easy to calculate: if today is 4 September, the deadline is the 24th. But what if today is 19 July?

An easy approach is to add in the usual manner and obtain 39 July. Since July has only 31 days, this must be converted by subtracting 31 to obtain 8 August.

Subtraction of days works similarly. Which day is 20 days before 8 August? Why, −12 August, of course! That is 31 − 12 July, or 19 July.

Another way is to subtract 8 days to obtain 0 August, which, being the day before 1 August, is 31 July. Since 20 − 8 = 12, we must now subtract 12 more days from 31 July to obtain 19 July.

Which day is 90 days after 24 September? Adding 30 days (the length of September) takes us to 24 October; 31 more days, to 24 November; 30 more, to 24 December. That makes 91 days (30 + 31 + 30), which is one too many, so the answer is 23 December.

In a common year (*i.e.*, one that is not a leap year), which day is 90 days before 5 March? Going back 28 days (the length of February) takes us to 5 February; 31 more days, to 5 January; 31 more, for a total of 90, to 5 December of the previous year.

Which day is 8 weeks after 29 March? We can add $8 \cdot 7 = 56$ days, getting 85 March. Subtracting 31 days for March and 30 days for April leaves 24, so the date is 24 May. Another way is to start by moving one week forward, into the next month. Since 2 days of March remain after the 29th, moving one week forward takes us to $7 - 2 = 5$ April. Having added 1 week, we add 7 weeks more: $5 + 49 = 54$ April, which is $54 - 30 = 24$ May. To determine which day was 8 weeks before 24 May, we can subtract the 24 days in May and all 30 of April: $56 - 24 - 30 = 2$. Thus we subtract 2 days from the 31 days of March to obtain 29 March.

If dates such as 39 July and −12 August disturb you, consider that the arrangement of the months is a mere matter of convention and that the *social* unacceptability of such dates does not bar you from using them *privately* for efficient mental calculation. Most people could not perform the calculations in this section; they would count days on the fingers or hunt for a calendar. By being flexible and taking full advantage of the number system, you can achieve what others cannot.

Number of days between dates

You were on vacation from 9 May to 3 June. How long did your vacation last?

The answer depends on boundary conditions, the cause of many a mistake: do you include or exclude the endpoints? If you count both 9 May and 3 June as days of vacation, you would say that your vacation included all of the days in May after the 8th, and also the first 3 days of June. Since May is 31 days long, your vacation lasted $31 - 8 + 3 = 26$ days.

If instead you take the view that your vacation did not start on 9 May but only the day after, you would perform the calculation similarly, but with the 9th rather than the 8th as the point of reference, and would obtain 25 days as the answer. Likewise, if you considered 3 June not to be a day of vacation for whatever reason, you would count only 2 days in June and again would obtain a different result.

School ends on 15 June and resumes on 4 September. How many days of leisure do your children enjoy? All of June after the 15th (in other words, all of June but the first 15 days), which is $30 - 15 = 15$ days; all of July, 31 days; all of August, 31 days; and the first 3 days of September (school starts on the following day). Total: $15 + 31 + 31 + 3 = 80$ days. What an opportunity to study mental arithmetic!

An easy way to add the days in multiple months is to start with 30 days per month and then add 1 for any 31-day month and subtract 2 for February (1 in a leap year). In 2024, how many days are there between 7 January and 26 October, not counting the endpoints? We have $31 - 7 = 24$ days in January, 25 in October, and all of the days in the 8 months in between. Count $30 \cdot 8 = 240$ days minus 1 for February (2024 is a leap year) plus 4 for the extra day in March, May, July, and August. The total is $24 + 25 + 240 - 1 + 4 = 289 + 3 = 292$.

Calculating the day of the week

The regularity of the calendar enables us to calculate the day of the week for any date. Remember that the approach presented here is based on the Gregorian calendar and therefore may give incorrect results for periods, such as the time before 1752 in England, when the Gregorian calendar was not in use.

To calculate the day of the week, we shall maintain a running total, using the following information:

- A code for the century (the year with the last two digits changed to 00). The table below shows the codes—0, 6, 4, 2—for centuries from 1500 to 2200.
- The last two digits of the year.
- Whether the year is a leap year or not.

- A code for the month. See the table below.
- The day.

Step 1: Obtain the code for the century from this table (centuries that are not listed continue the pattern):

1500, 1900	0
1600, 2000	6
1700, 2100	4
1800, 2200	2

Step 2: Add the last two digits of the year.

Step 3: Divide the last two digits of the year by 4 and add the result to the running total. Ignore any remainder or fraction. For example, the year 2023 yields 5 ($23 \div 4$, without the remainder) for this step.

Step 4: If the year is a leap year *and* the month is January or February, subtract 1 from the running total (to reverse the adjustment for leap year that was added in the previous step).

Step 5: Add the code for the month from the following table:

January	0	April	6	July	6	October	0
February	3	May	1	August	2	November	3
March	3	June	4	September	5	December	5

Step 6: Add the day.

Step 7: Divide the running total by 7 to obtain the remainder. (Ignore the quotient.) Convert the remainder to the day of the week by means of this table:

0	Sunday	4	Thursday
1	Monday	5	Friday
2	Tuesday	6	Saturday
3	Wednesday		

A few examples will illustrate the technique.

Today is 11 February 2024. Let us prove that it is a Sunday:

Step 1: The century is 2000; its code is 6. Start the running total with 6. *Step 2:* Add 24 (for 2024), yielding 30. *Step 3:* Add 6 (24 ÷ 4), yielding 36. *Step 4:* Since 2024 is a leap year and the month is February, subtract 1, yielding 35. *Step 5:* Add 3 for February, yielding 38. *Step 6:* Add 11 (the day of the month), yielding 49. *Step 7:* The remainder from 49 ÷ 7 is 0, so 11 February 2024 is indeed a Sunday.

I was born on 11 April 1970. Which day of the week was that? We have 0 for the century (1900). Add 70 for the year, yielding 70. The year divided by 4 is 17 (ignore the remainder); the total stands at 87. There is no adjustment for the first two months of a leap year. Add 6 for April, yielding 93. Add 11 for the day, yielding 104. Division by 7 gives a remainder of 6, so I was born on a Saturday.

Which day of the week will 25 December 2121 be? Take 4 for the century; adding 21 makes 25; adding 5 (year divided by 4) makes 30; no adjustment for leap year; adding 5 for December makes 35; adding 25 for the day makes 60. The remainder of 60 ÷ 7 is 4, so the day will be a Thursday.

This method is easy to use on paper, but you can carry all of the calculations out in your head if you memorise the tables. (The table of days of the week need not be memorised: just remember that Sunday is 0 and the other days follow in sequence.) I think of the codes for the months in groups of three, by quarter of the year: 0 3 3, 6 1 4, 6 2 5, 0 3 5. The codes for the centuries cycle through the even numbers below 7 from highest to lowest: 6 4 2 0.

Because only the remainder, not the quotient, is used in the final step, any multiple of 7 may be discarded along the way. At step 4 of the first example, we could have seen that $35 = 7 \cdot 5$ and reduced the running total to 0. In the second example, there was no need to add 70 (which is $7 \cdot 10$). In the third example, we could have skipped the addition of 21, or we could have dropped the running total of 35. If we were finding a date in 1974 or 2074, we could omit the 70 and add only 4, rather than 74, for the year.

The codes for the centuries and months take 1900 as the default century, because I began to use this method for calculating dates in the early 1980s and therefore set it up for the dates that were then current. Having used it for more than forty years, I am reluctant to change my habit now. You, however, may make 2000 the default century, with a code of 0, by substituting 0 5 3 1 for the cycle of centuries and subtracting 1 from each code for a month (except that 0 becomes 6, not −1, for convenience' sake; the elimination of multiples of 7 makes 6 and −1 functionally equivalent). The months from January to December are then 6 2 2, 5 0 3, 5 1 4, 6 2 4. Since this adjustment amounts to adding 1 to the century and subtracting 1 from the month, the calculation comes out the same.

Chapter 12

Measurements

UNITS OF MEASUREMENT are a source of frustration, especially in those few parts of the world that have not yet fully adopted or implemented the metric system (Système international). Yet converting units is not difficult if one proceeds carefully. Using the methods in this chapter, one can convert even compound units in one's head.

I pine for the day when the antiquated, retrograde, unscientific units still maintained by metric-resistant Anglo countries will be consigned to the ashcan of history.[45] Even then, however, we shall have occasion to convert units. The simple method in this chapter works for all units that differ only by a scaling factor. Conversions that use both a scaling factor and an offset, such as those between degrees Celsius and degrees Fahrenheit, require a different approach.

45 People in these countries often do not know just how bad their units of measurement are. They casually speak of ounces, yet few of them know that the ounce is a unit of weight *and* a unit of liquid measure—actually, *three* units of weight (avoirdupois, apothecaries', and troy ounces) and *two* of liquid measure (Imperial and US ounces). The avoirdupois pound has 16 ounces; the apothecaries' and troy pounds, only 12 ounces each. There are long tons, short tons, register tons, displacement tons, and freight tons.

General approach to conversion

When quantities are equal, dividing one by the other yields 1; multiplying by their quotient makes no change. Therefore, we can convert units in the following way, using the fact that 1 day (d) has 24 hours (h):

$$\begin{aligned} 7\,\text{d} &= (7\,\cancel{\text{d}})\frac{24\,\text{h}}{1\,\cancel{\text{d}}} \\ &= (7 \cdot 24)\,\text{h} \\ &= 168\,\text{h}. \end{aligned}$$

And

$$\begin{aligned} 72\,\text{h} &= (72\,\cancel{\text{h}})\frac{1\,\text{d}}{24\,\cancel{\text{h}}} \\ &= \frac{72}{24}\,\text{d} \\ &= 3\,\text{d}. \end{aligned}$$

We contrive to place any unwanted unit in the numerator and the denominator so that it will cancel out. For the same reason, we ensure that any desired unit appears in the numerator or denominator, wherever it belongs.

A car moves at $10\overline{0}$ kilometres per hour (measured to three significant digits). What is its speed in metres per second? Set the work up as follows:

$$\frac{100\,\text{km}}{\text{h}} \cdot \frac{1000\,\text{m}}{\text{km}} \cdot \frac{\text{h}}{60\,\text{min}} \cdot \frac{\text{min}}{60\,\text{s}}.$$

Then

$$\begin{aligned} \frac{100\,\cancel{\text{km}}}{\cancel{\text{h}}} \cdot \frac{1000\,\text{m}}{\cancel{\text{km}}} \cdot \frac{\cancel{\text{h}}}{60\,\cancel{\text{min}}} \cdot \frac{\cancel{\text{min}}}{60\,\text{s}} &= \frac{100 \cdot 1000\,\text{m}}{60 \cdot 60\,\text{s}} \\ &= \frac{1000\,\text{m}}{36\,\text{s}} \end{aligned}$$

$$
\begin{aligned}
&= \frac{250\,\mathrm{m}}{9\,\mathrm{s}} \\
&= 27.\overline{7}\,\frac{\mathrm{m}}{\mathrm{s}}.
\end{aligned}
$$

To three significant digits, the answer is 27.8 m/s.

A Canadian in the US sees that gasoline costs 3.24 US dollars (USD) per US gallon. To compare that price to those at home, where they are given in Canadian dollars (CAD) per litre, she needs help with conversion. At the time of writing, a US dollar is worth $1.35 Canadian.

$$
\begin{aligned}
3.24\,\frac{\mathrm{USD}}{\mathrm{gal}} &= \frac{3.24\,\cancel{\mathrm{USD}}}{\cancel{\mathrm{gal}}} \cdot \frac{1.35\,\mathrm{CAD}}{\cancel{\mathrm{USD}}} \cdot \frac{\cancel{\mathrm{gal}}}{3.785\,\mathrm{l}} \\
&= \frac{3.24 \cdot 1.35\,\mathrm{CAD}}{3.785\,\mathrm{l}} \\
&= \frac{3.24 \cdot 1.35 \cdot 4\,\mathrm{CAD}}{3.785 \cdot 4\,\mathrm{l}} \\
&= \frac{3.24 \cdot 5.4\,\mathrm{CAD}}{15.14\,\mathrm{l}} \\
&\approx \frac{17.5\,\mathrm{CAD}}{15(1 + 1\%)\,\mathrm{l}} \\
&\approx \left(\frac{17.5}{15} - 1\%\right)\frac{\mathrm{CAD}}{\mathrm{l}} \\
&\approx (1.17 - 0.01)\frac{\mathrm{CAD}}{\mathrm{l}} \\
&= 1.16\,\frac{\mathrm{CAD}}{\mathrm{l}}.
\end{aligned}
$$

For the approximation near the end, I calculated $0.324 \cdot 9 \cdot 6 = 17.496$ (moving the decimal point and factoring 54). The original figure was given to only three significant digits, so I rounded 17.496 to 17.5. Ordinarily I would not correct for significant digits in the middle of a calculation, but 17.496 was so close to 17.5 that rounding was safe.

Currency

Travellers, buyers, and sellers often need a rough estimate of the value of an amount of money in a different currency. Mental arithmetic cannot help with the fluctuation of exchange rates, but it can provide ways of estimating conversions quickly and accurately. With just a bit of preparation, you can develop a simple method of estimation that will work as long as the exchange rate remains fairly stable. A few examples will illustrate the approach.

Suppose that you live in Canada and are planning to travel around the world. Wherever you go, you wish to be able to estimate the equivalent in Canadian dollars of any price. You therefore need rules for approximate conversion that require only one or two easy calculations. You can tolerate an error of 1% or so.

At the time of writing, one Canadian dollar was worth 1.119 1 Australian dollars. You can therefore convert an Australian price to Canadian by dividing by 1.119 1. Since that is inconvenient, note that 1.119 1 is approximately 1.111 1, or $\frac{10}{9}$. Dividing by $\frac{10}{9}$ is the same as multiplying by its reciprocal, $\frac{9}{10}$, which in turn is the same as subtracting 10%. Thus a meal that costs \$32 Australian is about \$29 Canadian—a little less, since the divisor was reduced.

There are 67.486 Russian roubles to the Canadian dollar. Instead of dividing by this unwieldy figure, you note that it is about $\frac{2}{3} \cdot 100$; you can therefore multiply by $\frac{3}{2}$ and divide by 100. Multiplying by $\frac{3}{2}$ is the same as adding half; thus your rule is to add half to the price in roubles and divide by 100. A hotel room at 3 200 roubles costs about $32 + 16 = 48$ Canadian dollars.

One dollar is worth 266.59 Hungarian forints. This figure is close to $266.66 \approx \left(2 + \frac{2}{3}\right)100 = \frac{8}{3} \cdot 100$; dividing by it, therefore, amounts to dividing by 100 and multiplying by $\frac{3}{8}$. (Another way to see the same thing is to calculate its reciprocal: $\frac{1}{266.59} \approx 0.003\,75$, which is recognisable as $\frac{3}{8} \cdot \frac{1}{100}$.) It may be easier, however, to divide by 200 and subtract $\frac{1}{4}$ of the result (thereby obtaining $\frac{3}{4} \cdot \frac{1}{200} = \frac{3}{8} \cdot \frac{1}{100}$). A 67 000-forint bill is about $335 - 85 = 250$ Canadian.

Unlike the other units of currency mentioned in this section, the British pound sterling is worth more than a Canadian dollar: conversion requires not division but multiplication, by 1.739 4. This figure is so close to 1.75 that it suggests the rule of adding $\frac{3}{4}$ of the amount. Easier, however, is to double the amount and subtract $\frac{1}{8}$ of the result (because $2 - 2 \cdot \frac{1}{8} = 2 - \frac{1}{4} = 1.75$). A £15 purchase, therefore, costs about $30 - 4 = 26$ Canadian.

The Chinese exchange rate is less convenient, at 5.312 3 yuán to the dollar. You could simply divide by 5 and accept the resulting inaccuracy. Instead, you might compute the reciprocal $\frac{1}{5.312\,3} = 0.1882$ and see whether it suggested a better approximation. Since it is approximately 19%, you could adopt the rule of taking 20% of the figure in yuán (which is the same as dividing it by 5) and subtracting 5% of it (because $20\% - (20\% \cdot 5\%) = 20\% - 1\% = 19\%$). In the alternative, since $0.1882 = 0.1667 + 0.0215 \approx 0.1667 + \frac{1}{8} \cdot 0.1667$, you could divide by 6 (multiply by 0.1667) and add $\frac{1}{8}$ of the result. Using the first technique, a 490-yuán purchase costs about $98 - 5 = 93$ Canadian; using the second, it is $82 + 10 = 92$. (The actual Canadian value is $92.24.)

When travelling, we can sometimes use one local banknote as a benchmark for a convenient amount of our own currency, thereby obtaining an easy, if approximate, way to think directly in local currency. For instance, one Canadian dollar is worth 12.419 Mexican pesos. It might be helpful to think of the 100-peso banknote as an 8-dollar bill; then a 600-peso transaction would cost about $48, and 50 pesos would be about $4.

Fancier conversions

Men's hat sizes in North America are measured in inches of diameter, calculated from a measurement of the skull above the eyebrows. (The head certainly is not circular at that site or any other; the periphery is taken to be the «circumference», and the «diameter» is determined accordingly.) In Europe and many other parts of the world, centimetres of circumference are used.

My European hat size is 61. What is the North American equivalent? We start by converting centimetres of circumference to inches of circumference:

$$
\begin{aligned}
(61\,\cancel{\text{cm}})\frac{\text{in}}{2.54\,\cancel{\text{cm}}} &= \frac{61}{2.54}\,\text{in} \\
&\approx \frac{61 - 1.6\%}{2.5}\,\text{in} \\
&\approx \frac{60}{2.5}\,\text{in} \\
&= \frac{240}{10}\,\text{in} \\
&= 24\,\text{in}.
\end{aligned}
$$

Then we convert 24 inches of circumference to inches of diameter. Since the circumference is π times the diameter, we divide by π, taking advantage of the fact, demonstrated in the section on fractions and decimals, that $32\pi \approx 100.48 \approx 100 + \frac{1}{2}\%$.

$$
\begin{aligned}
\frac{24}{\pi} &= \frac{24 \cdot 32}{32\pi} \\
&\approx \frac{768}{100.48} \\
&\approx \frac{768 - \frac{1}{2}\%}{100} \\
&\approx \frac{764}{100} \\
&= 7.64.
\end{aligned}
$$

My North American hat size is therefore about 7.64. Since these sizes are given to the nearest eighth, we substitute $7.625 = 7\frac{5}{8}$.

A few conversions require the use of reciprocals. The fuel efficiency of motor vehicles in the US is measured in miles per US gallon; thus a large value is desirable. Most of the world measures fuel efficiency in litres per 100 kilometres; thus a small value is desirable. The general method of this chapter cannot be used directly, because miles in the numerator will not turn into kilometres in the denominator. Since the US measures distance divided by volume while most of the rest of the world measures volume divided by distance, the method will work if we begin with the *reciprocal* of the measurement being converted.

A car in the US gets 32 miles per gallon. What is that in litres per 100 kilometres?

$$\begin{aligned}
\frac{\cancel{\text{gal}}}{32\,\cancel{\text{mi}}} \cdot \frac{3.785\,\text{l}}{\cancel{\text{gal}}} \cdot \frac{\cancel{\text{mi}}}{1.609\,\cancel{\text{km}}} \cdot \frac{100\,\cancel{\text{km}}}{100\,\text{km}} &= \frac{3.785 \cdot 100\,\text{l}}{(32 \cdot 1.609)(100\,\text{km})} \\
&\approx \frac{11.83\,\text{l}}{(1.609)(100\,\text{km})} \\
&\approx \frac{11.83\,\text{l}}{\left(1.6(1 + \frac{1}{2}\%)\right)(100\,\text{km})} \\
&\approx \frac{11.83\left(1 - \frac{1}{2}\%\right)\,\text{l}}{1.6(100\,\text{km})} \\
&\approx \frac{(11.83 - 0.06)\,\text{l}}{1.6(100\,\text{km})} \\
&= \frac{11.77\,\text{l}}{1.6(100\,\text{km})} \\
&\approx 7.4\,\frac{\text{l}}{100\,\text{km}}.
\end{aligned}$$

This answer is correct, to two digits (the accuracy of the original measurement).

Multiple units

Cruel Mr Murdstone, aided and abetted by his equally cruel sister, put the following problem to poor young David Copperfield: «If I go into a cheesemonger's shop, and buy five thousand double-Gloucester cheeses at fourpence-halfpenny each, present payment.»[46]

I wish that David had wryly observed that no retail cheesemonger could deliver five thousand double-Gloucester cheeses and that Mr Murdstone would have had no use for that many. Instead, being much harassed by the Murdstone pair, he failed to answer the question and was given for dinner nothing but one slice of bread.

With our modern decimal currency, multiplying $4\frac{1}{2}$ by 5 000 would be easy, even for a child languishing under the oppression of a wicked stepfather. In David's day, however, Britain used a clumsy currency in which there were 4 farthings to the penny, 12 pence to the shilling, and 20 shillings to the pound.[47] This piece of folly persisted until 1971, when the pound was redefined to have 100 (new) pence and the other units were abolished.

Dear David deserves our help. First, we calculate the amount in pence: $5\,000 \cdot 4 = 20\,000$, and $5\,000 \cdot \frac{1}{2} = 2\,500$; the bill therefore comes to 22 500 pence. Since the cheesemonger does not want coins enough to cause the floor beneath the till to cave in, let us convert that to shillings: $22\,500 \div 12 = 1\,875$. Because there is no remainder, no part of the payment need be made in pence. Now we convert to pounds: $1\,875 \div 20 = 187.5 \div 2 = 93.75$, which is 93 remainder $\frac{3}{4} \cdot 20 = 15$. Whisper to David that the answer is £93/15 (93 pounds 15 shillings).

46 Charles Dickens, *The Personal History of David Copperfield* (London: Bradbury & Evans, 1850), c 4 at 40–41. The novel is partly autobiographical; perhaps Dickens as a boy was tormented with a similar problem.

47 As if that were not enough, lawyers collected 5% barristerial vigorish by charging their fees in guineas, a guinea being a pound and a shilling.

When I was in school, arithmetic textbooks featured problems such as this: «Mary has 17 yards, 2 feet, and $11\frac{7}{8}$ inches of cloth. She wishes to cut it into pieces measuring 1 yard, 1 foot, and $8\frac{19}{32}$ inches. How many pieces can she make, and how much cloth will be left over?»

I could only wonder at the stupidity of the problem: *If Mary had any sense, she would have measured her cloth in a single unit —preferably metres or centimetres, but even inches would do—and spared herself a lot of tedious arithmetic.* You can imagine how popular I was with the teachers.

My smart-alecky observation, however, pointed the way forward: convert everything to inches. Since there are 3 feet in a yard, 17 yards are $17 \cdot 3 = 51$ feet; then, there being 12 inches in a foot, we get $(51 + 2) \cdot 12 + 11\frac{7}{8} = 53 \cdot 12 + 11\frac{7}{8} = 636 + 11\frac{7}{8} = 647\frac{7}{8}$ inches. Similarly, the length of each cut piece will be $56\frac{19}{32}$ inches. To calculate $647\frac{7}{8} \div 56\frac{19}{32}$, we can multiply dividend and divisor by 32 to eliminate the fractions: $20\,732 \div 1\,811$. The answer is 11 whole pieces of cloth, with a remainder of 811 that represents thirty-seconds of an inch (because we multiplied by 32 before dividing). The cloth left over measures $811 \div 32 = 25\frac{11}{32}$ inches. Since 24 inches make 2 feet, we convert this remainder to 2 feet and $1\frac{11}{32}$ inches. Thus Mary can cut 11 pieces and will have 2 feet and $1\frac{11}{32}$ inches of cloth left over.

I hope that you will not have to perform this sort of calculation ever again, especially in your head.

Much easier is division by a constant. Suppose that Mary had wanted to cut her cloth into 5 pieces. How long would they be?

We can begin by dividing 17 yards by 5, obtaining 3 remainder 2. The remainder, 2 yards, we convert to feet: $2 \cdot 3 = 6$ feet. We then add these 6 feet to the 2 feet of Mary's measurement to obtain 8 feet, which, divided by 5, yields 1 remainder 3. Again, 3 feet is $3 \cdot 12 = 36$ inches, which we add to $11\frac{7}{8}$ for a total of $47\frac{7}{8}$ inches. Dividing by 5 yields 9 remainder $2\frac{7}{8}$. Therefore, each piece of cloth will measure 3 yards, 1 foot, and 9 inches, with $2\frac{7}{8}$ inches remaining—which we can convert to $\frac{8 \cdot 2+7}{8} = \frac{23}{8}$; divided by 5, that is exactly $\frac{23}{40}$ of an inch per piece of cloth, so for the inches we have not 9 but a bit more than $9\frac{1}{2}$. Unlike the previous problem, this one can be solved mentally without weeping or gnashing of teeth.

Chapter 13

Remembering numbers

BY FAR the greatest challenge for the mental calculator lies in retaining numbers in memory. Proceeding from left to right, announcing digits in the answer as they are found, and employing efficient algorithms can reduce, but not eliminate, the need to remember intermediate results. Even the most skilful people find themselves forgetting digits, especially in the midst of calculation. With little context to make them memorable, numbers are easily forgotten.

A useful technique for remembering numbers, whether temporarily or permanently, is the *mnemonic major system*. With a knowledge of the system and a bit of creativity, one can convert undifferentiated strings of digits into memorable words, from which the digits can be unambiguously recovered.

Each digit is associated with one or more sounds, shown in the table below with mnemonic aids for learning the system.[48]

Unassigned are all vowels and the consonants *h*, *w*, *wh*, and *y*. These may be used freely to make words; they have no numeric value. Thus the words *highway* and *yo-yo* do not correspond to any number.

---

48 Because the mnemonic major system was designed for the French language, it originally made no provision for the sounds of *ng* and *th*, which do not occur in native French words. The assignments of these sounds in the table reflect my preferences; other authors treat them differently.

0	s, z	*Zero* begins with *z*
1	t, d, th	One downstroke in *t* and *d*
2	n, ng	Two downstrokes in *n*
3	m	Three downstrokes in *m*
4	r	*Four* ends in *r*
5	l	Spread five fingers; index & thumb form *L*
6	sh, ch, j, zh	*6* in mirror image resembles a script *j*
7	k, g	*K* can be made from two *7*s
8	f, v	*8* resembles a script *f*
9	p, b	*9* in mirror image resembles *P*

Note that the table lists sounds, not spellings. *Gem* begins with a *j* sound; it is therefore interpreted as *j-m* (63), not as *g-m* (73). Likewise, *laugh* is *l-f* (58), not *l-g* (57); *rice* is *r-s* (40); *comb*, with its silent *b*, is *k-m* (73), not *k-m-b* (739); and *bottle* is *b-t-l* (915), not *b-t-t-l* (9115).

Consonants that occur in both voiced and voiceless forms are conveniently assigned to the same digit: s/z, t/d, th (voiceless as in *thigh*, voiced as in *thy*), sh/zh, ch/j, k/g, f/v, p/b.

To remember a number, form a memorable word or phrase from the sounds that its digits represent. It is easiest to think of only one sound per digit; the other sounds are closely related, and with practice you will be able to think of all of them when you think of one. Remember that you may add vowels, *h*, *w*, *wh*, and *y* as you please, since they have no numeric value.

For example, suppose that you had read a book up to page 94. Lacking a bookmark, you needed to remember the page number. Since 94 corresponds to *p-r* (and to *b-r*), you could associate it with any of the following words, listed here in the order in which they occurred to me: *bear*, *pear*, *beer*, *berry*, *poor*, *burrow*, *bar*, *opera*, *apiary*, *wiper*, *bury*, *purée*, *beret*, *power*, *brow*, *Peru*. (Many others exist.) Of these, the most memorable choices are nouns that are easy to visualise: *bear* may be the best, but *pear* and *beer* are also good; *poor*, *burrow*, and *bury* are weak. All that matters is that the word be memorable to you, since only you will use it.[49] *Opera* is memorable to me, fond as I am of the lyric art; to a beekeeper, *apiary* might stand out better.

Having chosen your word, you then associate it with what you are trying to remember—the position in your book—in a way that is hard to forget. Fanciful, even silly, images tend to be more effective than bland ones. If you chose *bear* (as I would), you might imagine opening your book to see a bear lunge at you from the pages, with claws and teeth bared; or perhaps a bear was avidly reading your book and you had to wrest it out of the bear's paws. For *beer*, you might see the book soaking in a filled stein, or stinking so strongly of stale beer that you can scarcely bear to read it. For *wiper*, you might see yourself turning on windshield wipers so that you could read the rain-drenched pages. For *opera*, you might be shocked to hear the sextet «Chi mi frena in tal momento?» from *Lucia di Lammermoor* fill the room as soon as you opened the book.

When you return to your book, the word will come to mind, and you can easily convert it back to the page number.

Suppose that you would like to remember that my birthday is April 11. Converting 411 to a word requires a bit of thought: *hardhat* comes to mind, along with *redhead* and a few more.[50] The image of me wearing a hardhat in front of a blackboard covered with calculations could cement the date in your memory. If

---

49 In *The Memory Book* at 117, the late Harry Lorayne recalled a contest in which he came up with *somevere* as a word for 0384: a belovèd neighbour from his childhood pronounced *somewhere* in that idiosyncratic way. Noöne else would have thought of that or found it meaningful. When his co-author, Jerry Lucas, suggested *samovar*, Lorayne admitted that he did not know the word. I would have used *samovar* or *semaphore*; I can readily envision both, but perhaps Lorayne would have found *semaphore* equally obscure.

50 Some numbers, especially those with repeated digits, correspond to fewer words than others. Repetition of 8, 6, or 4 can be especially inconvenient, at least if the block of repeated digits cannot be split across words.

you struggled to think of a single word, you might try a phrase, such as *hairy teddy*, *wiry teeth*, or *red tie*. Since you know that the date cannot be longer than three digits, you could even add extraneous sounds to form *radiator* (4114) or *red tape* (4119); when reconstructing the date, you would drop the unneeded digits.

To remember a list of items in order, you can set up a list of *peg words* that correspond to numbers, using the mnemonic major system. Choose those that suit you best, but for illustration we shall use for 1 to 10 *hat*, *hen*, *ham*, *hair*, *whale*, *shoe*, *cow*, *ivy*, *bee*, and *daisy*. You can extend the set to 100, even to 1 000 if necessary.

Suppose that today you need to do three things in order: mail a letter, pick up the dry cleaning, and buy wine. For item 1, imagine putting a stamp on your *hat* and stuffing the hat into a mailbox, or having your mail delivered in an upturned hat. For item 2, imagine a *hen* wearing your suit. For item 3, baste a *ham* with wine, or fill your glass from a ham gushing with Cabernet Sauvignon. To recall the tasks in order, simply think of your peg words; the associations will come to mind.

You could use the same technique to remember the chemical elements by atomic number. Carbon is element 6: think of having an enormous *shoe* as your *car* (similar in sound to *carbon*). Neon is 10: see a field of *daisies* glowing like a neon sign. Silver is 47: think of polishing a silver *rake* (47: *r-k*) before going outside to rake the leaves in style. Osmium is 76: think of the Wizard of *Oz* in a *cage* (76: *k-j*).

For public speaking, key the topics of your speech to your peg words. You may then be able to speak without notes.

Other arbitrary numbers can be memorised by association. How to remember that Mozart's Clarinet Quintet is K. 581? See it performed on a *lift* (elevator), or in a *loft*. Herbert Hoover was the 31st president of the US: think of a *maid* with a *Hoover* vacuum cleaner. The departments (administrative divisions) of France are officially numbered, and Hérault is 34: perhaps the *mayor* is a *hero*. With a little creativity and effort, you can memorise telephone numbers or any other numbers that you may need.

For short-term memory, as when performing mental arithmetic, it suffices to convert the number to a word or phrase that can be remembered for a moment; there is no need to devise an enduring image for a number that will be discarded soon enough.

Well practised, the mnemonic major system facilitates astonishing feats of memory. A rich literature on memory discusses its broad application to remembering everything from numeric data to card games to ordered lists of information. Henceforth in this book, however, it is used only as a mnemonic aid for short-term recall of numbers during mental calculation. Examples of its use appear in subsequent chapters.

Chapter 14

Squares of four-digit and larger numbers

HOWEVER SIMILAR they may seem to the two- and three-digit squares studied in a previous chapter, squares of numbers with four or more digits are best approached with techniques of their own, since the need to remember and combine intermediate results increases the calculator's burden. This chapter presents practical ways to square large numbers without overtaxing the memory.

General method for four-digit numbers

The technique for three-digit numbers works well for four-digit numbers as long as one can remember a few extra digits. The following example looks more fearsome than it is:

$$\begin{aligned}
6\,283^2 &= 6\,000 \cdot 6\,566 + 283^2 \\
&= 39\,396\,000 + 283^2 \\
&= 39\,396\,000 + 300 \cdot 266 + 17^2 \\
&= 39\,396\,000 + 79\,800 + 289 \\
&= 39\,396\,000 + 80\,089 \\
&= 39\,476\,089.
\end{aligned}$$

For mental calculation, we can simplify the work in the following manner:

First, double 283 to obtain 566. (As usual, keep only the last three digits and disregard any 1 in the thousands' place.) Then $6 \cdot 6\,566$ will give the number of millions in the answer. Multiply in pairs ($6 \cdot 65 = 390$; $6 \cdot 66 = 396$; combine to make 39 396) or digit by digit ([36, 30, 36, 36] = 39 396). The square is therefore 39 396 000 plus 283^2.

To relieve the burden of keeping track of the millions, and also to impress others with the speed of our mental arithmetic, we can announce the number of millions as soon as we know it. The only question is whether the final step of the calculation (here, 283^2) will cause the carrying of 1 into the millions' place. Here we know that there are either 39 or 40 millions. Once we have resolved this question, we can begin to declare the answer.

If the number of thousands already calculated is less than 750, there will be no carrying into the millions' place, because the square to be added is no greater than $500^2 = 250\,000$. In our example, with 396 thousands, we see that there is nothing to carry; we therefore announce «39 million ...» as the first part of the answer.

We must now make sure to remember 396 while we compute 283^2. I find that I can maintain the figure in short-term memory, but many people cannot. The safest course is to deploy the mnemonic major system by converting 396 to a word or phrase. Its digits represent *m-p-sh*, from which we might form *impeach*, *ambush*, *impish*, *my badge*, *hemp shoe*, or something else.

Once we have encoded 396, we can forget about it and proceed with $283^2 = 300 \cdot 266 + 17^2 = 79\,800 + 289 = 80\,089$. This we treat as 80 thousands plus 89.

Finally, we recall the memorised value 396, if necessary by converting *ambush* back to numeric form. Since 396 represents the number of thousands from the first part of the calculation, we add 80 to obtain 476. We can then announce the rest of the answer: «... 476 thousand 89» (39 476 089).

Another example: to square 7 847, we double 847 to obtain 694 (disregarding the carried 1 as usual). Then we multiply $8 \cdot 7\,694 = 8(7\,700 - 6) = 61\,600 - 48 = 61\,552$. Since $552 < 750$, there will be no carrying into the millions, and we can announce «61 million ...». If necessary, we convert 552 (*l-l-n*) to *Lillian* or *lulling* or *low line*. For the final step, we square $1000 - 847 = 153$ to obtain $200 \cdot 106 + 47^2 = 21\,200 + 2\,209 = 23\,409$. Adding $552\,000 + 23\,409$ gives 575 409, which we announce to complete the answer (61 575 409).

To compute $3\,861^2$, we start with $4 \cdot 3\,722 = 14\,888$. Because $888 > 750$, we are not yet certain that we have 14 rather than 15 millions. A bit of analysis shows that 139^2, the amount that we are adding, is around 20 000 ($140^2 = 19\,600$) and therefore is not large enough to cause carrying when added to 888 thousand. Consequently, we can announce «14 million ...». Perhaps 888 is distinctive enough to be remembered without assistance, but if necessary we can convert it to *vivify* or *heavy fife*. Then $139^2 = 100 \cdot 178 + 39^2 = 17\,800 + 1\,521 = 19\,321$. That makes $888 + 19 = 907$ thousands and 321, or «... 907 thousand 321» (14 907 321).

Alternate method

An alternate method works well for special cases, notably those whose second or third digit is 0 (sometimes 5) and those that are only slightly lower than a convenient round number. The method takes advantage of the facts that

$$(a+b)^2 = a^2 + 2ab + b^2.$$

and

$$(a-b)^2 = a^2 - 2ab + b^2.$$

Although $6\,207^2 = 6\,414 \cdot 6\,000 + 207^2$, it is more easily computed as $(6\,200 + 7)^2 = 6\,200^2 + 6\,200 \cdot 7 \cdot 2 + 7^2$. First, since $62^2 = 3\,844$, we have $6\,200^2 = 38\,440\,000$. No millions will be carried, so we can announce «38 million ...». Next, we must calculate $6\,200 \cdot 7 \cdot 2$. With a knowledge of the multiples of 14, we can get $62 \cdot 14 = [84, 28] = 868$ very easily; without that knowledge, it is easiest to multiply first by 7 and then double: $62 \cdot 7 \cdot 2 = 434 \cdot 2 = 868$. Since we omitted the two zeroes from 6 200, the product is not 868 but 86 800, which we add to 440 000 to get 526 800.[51] The rest of the answer is simply $7^2 = 49$; thus $6\,207^2 = 38\,526\,849$.

The usual method works well for 4 392, but we can take advantage of the fact that $4\,392 = 4\,400 - 8$. First we obtain $4\,400^2 = 19\,360\,000$ from a good knowledge of the squares. Then we compute $4\,400 \cdot 8 \cdot 2 = 35\,200 \cdot 2 = 70\,400$, or, directly, $4\,400 \cdot 16 = [64, 64]00 = 70\,400$; this must be subtracted, yielding 19 289 600 (using the complement: $[36-8, 96]$ or $[36-10, 296]$). Finally, we add $8^2 = 64$ to obtain 19 289 664.

The same method can be applied when the second digit of the four-digit number is 0. For instance, $7\,029^2 = (7\,000+29)^2$. We start with $7\,000^2$, which is 49 million. This we can immediately announce as the start of the answer if we perceive that $7\,000 \cdot 29 \cdot 2$ is less than 1 million and therefore will not cause carrying. (We have roughly $7 \cdot 30 \cdot 2 = 210 \cdot 2 = 420$ as the number of thousands, a number well below a million.) For that calculation, we first obtain $29 \cdot 7 = 203$, then double that result to 406; since we omitted the three zeroes from 7 000, we have not 406 but 406 000. Finally, $29^2 = 841$, and the answer is 49 406 841.

51 If you know the squares, you need not convert 440 using the mnemonic major system, since it comes from the last two digits of 62^2 and can be recalled easily. Even if you have to compute 62^2, you are likely to remember it until you need it. If you must, convert 44 to *rower*, *roar*, or *horror* and append a zero.

Numbers near 5 000

Squaring numbers near 5 000, especially those between 4 500 and 5 500, is easy: the first two digits minus 25 give the millions; the last two digits form the first two digits of the thousands; adding the square of the difference from 5 000 completes the work. A few examples will illustrate this useful shortcut.

Suppose that we want $5\,043^2$. We can break the number into two-digit components, 50 and 43. The number of millions is $50 - 25 = 25$; we announce «25 million ...». The last two digits, 43, form the next two digits of the answer; therefore, we have 430 thousand plus 43^2. Since $43^2 = 1\,849$, the rest of the answer is 431 849. Thus $5\,043^2 = 25\,431\,849$. If we know 43^2, our only calculation is $50 - 25$!

In fact, we need not subtract 25 from the first two digits: we can instead add to 25 the difference of the first two digits from 50. In the case of 5 043, or any other number that starts with 50, the difference is 0; therefore, our first two digits are $25 + 0 = 25$. For a number beginning with 52, our digits would be $25 + 2 = 27$. For a number beginning with 46, we would have $25 - 4 = 21$.

For $4\,762^2$, we have $25 - 3 = 22$ millions, plus $62 \cdot 10 = 620$ thousands, plus $(1\,000 - 762)^2 = 238^2$. We announce «22 million ...», then find $238^2 = 276 \cdot 200 + 38^2 = 55\,200 + 1\,444 = 56\,644$. The answer ends with $620\,000 + 56\,644 = 676\,644$, which we announce. The answer is 22 676 644.

Rarely do we have to carry into the millions. For instance, $5\,387^2$ seems to start as 28 million, but 870 000 added to $387^2 = 374 \cdot 400 + 13^2 = 149\,600 + 169 = 149\,769$ yields 1 019 769, for a total of 29 019 769. If the last two digits of the number being squared are close to 100, it may be good policy to check for carrying, postpone the announcement of the millions until their value is certain, or be prepared to correct the millions by adding 1.

Numbers near 10 000

Begin by doubling any number near 10 000 and disregarding the leading 1. The first two digits will give the millions, though carrying will occasionally increase the value. The last two digits of the doubled amount will begin the thousands. The square of the difference of the original number from 10 000 completes the calculation.

Thus $9\,817^2$ begins with $9\,817 \cdot 2 \rightarrow 9634$, and we can announce «96 million ...». To 340 000 we add $(1\,000 - 817)^2 = 183^2 = 166 \cdot 200 + 17^2 = 33\,200 + 289 = 33\,489$ to obtain 373 489. The answer in full is 96 373 489.

Numbers that are quite far from 10 000 can also be squared in this way. For $9\,146^2$, we calculate $9\,146 \cdot 2 \rightarrow 8\,292$ and add 82 920 000 to $(1\,000 - 146)^2 = 854^2$. Obviously there will be carrying into the millions, so we add 1 and announce «83 million ...». Then $854^2 = 808 \cdot 900 + 46^2 = 727\,200 + 2\,116 = 729\,316$, and the answer is $82\,920\,000 + 729\,316 = 83\,649\,316$.

Numbers somewhat in excess of 10 000 will have nine-digit squares. Double the excess, not the leading 1, and add the result to 100 million. For $10\,053^2$, we have $101\,060\,000 + 53^2 = 101\,062\,809$. Rarely, a larger number may require carrying into the millions: $10\,395^2 = 107\,900\,000 + 395^2 = 107\,900\,000 + 156\,025 = 108\,056\,025$.

Numbers of five or more digits

The general method based on the difference of squares does not work well for larger numbers, as the burden of remembering the intermediate results becomes prohibitively great and we still find ourselves calculating a four-digit square by the same means. Better in most cases is to divide the number into two parts: the thousands (1 000a) and the rest (b). We can calculate the square as

$$(1\,000a + b)^2 = 1\,000\,000a^2 + 1\,000(2ab) + b^2.$$

Take $71\,238^2 = (71\,000 + 238)^2$. The number of millions is $71^2 = 5\,041$ plus whatever is carried from the rest of the calculation. Already we can announce «5 billion ...» and retain 041 (the remaining millions) in memory, perhaps by converting it (*s-r-t*) to *sword*. (The initial zero could instead be dropped, leaving 41, which is *r-t*—*rat* or *heart* would be good.)

The next step is usually the most difficult: $2 \cdot 71 \cdot 238$, to obtain the thousands. We can begin by doubling a number, or we can leave doubling to the end. Doubling now does not seem advantageous (it might if, for example, one number contained a 5), so instead we compute $71 \cdot 238 = 70 \cdot 238 + 1 \cdot 238 = 16\,660 + 238 = 16\,898$. Doubling this figure yields 33 796, the number of thousands. Will the final step, 238^2, cause carrying into the millions? No, because 238^2 is only about 50 000, and $796{+}50 < 1\,000$. Therefore, the number of millions is $041{+}33 = 074$. We continue the answer by reciting «... 74 million ...», and we keep 796 in mind as *cabbage* (*k-p-j*).

Finally, we calculate $238^2 = 200 \cdot 276 + 38^2 = 55\,200 + 1\,444 = 56\,644$. Adding 56 to 796 left over from the previous step yields 852, and we complete the answer: «... 852 thousand 644.» We have just calculated $71\,238^2 = 5\,074\,852\,644$.

Another example: $63\,277^2$. We start with $63^2 = 3\,969$. If the next step of the calculation ($2{\cdot}63{\cdot}277$) yields 31 000 or more, there will be carrying into the billions' place, giving 4 billions instead of 3. Since it is hard to decide that question at a glance, we continue the calculation without announcing the billions: $2{\cdot}63{\cdot}277 = 2{\cdot}7{\cdot}9{\cdot}277 = 2{\cdot}7{\cdot}2493 = 2{\cdot}17\,451 = 34\,902$. (Another way: since $2 \cdot 277 = 554$, we can calculate $63 \cdot 555 - 63 = 315 \cdot 111 - 63$. Then $315 \cdot 111$ is simply 315 repeated three times, each time staggered one more place to the right. We can calculate it in two steps: $3\,150+315 = 3\,465$, then $34\,650+315 = 34\,965$. Deducting 63 yields 34 902.) Adding 34 to 3 969 gives 4 003; we announce «4 billion 3 million ...» and remember 902 as *bassoon*. (Carrying will not occur: 277^2 is well below $300^2 = 90\,000$, which would not cause carrying when added to 902 000.) Finally, $277^2 = 300 \cdot 254 + 23^2 = 76\,200 + 529 = 76\,729$. We add 76 to 902 and finish the answer: «... 978 thousand 729». In figures: $63\,277^2 = 4\,003\,978\,729$.

For a final display of legerdemain, let us attempt the square of a six-digit number, 258 619. Our strategy: $258\,619^2 = 258\,000^2 + 258\,000 \cdot 619 \cdot 2 + 619^2$.

First, we compute $258^2 = 300 \cdot 216 + 42^2 = 64\,800 + 1\,764 = 66\,564$. The next step will not cause carrying into the billions because its value is only about 300 000 ($2 \cdot 258 \cdot 619 \approx \frac{1}{2} \cdot 1\,000 \cdot 619$); therefore, we announce «66 billion ...» and retain 564 as *wheelchair* or *ledger*.

Next comes the hard part: $2 \cdot 258 \cdot 619$. It may be easiest to double 258 first, obtaining a 5 followed by something small. Here we calculate $16 \cdot 619$ as $16 \cdot 600 + 16 \cdot 19$; the $16 \cdot 19$ we calculate as $(20-4)(20-1)$, using 20 as a point of reference.

$$
\begin{aligned}
2 \cdot 258 \cdot 619 &= 516 \cdot 619 \\
&= 500 \cdot 619 + 16 \cdot 619 \\
&= \tfrac{1}{2} \cdot 619\,000 + 16 \cdot 619 \\
&= 309\,500 + 9\,600 + 16 \cdot 19 \\
&= 319\,100 + 20 \cdot 15 + 1 \cdot 4 \\
&= 319\,100 + 304 \\
&= 319\,404.
\end{aligned}
$$

Retain 404 as *razor*; compute 564 + 319 = 883 for the millions. There will be no millions to carry, since $619^2 < 400\,000$, which will not cause carrying when added to 404 000. Now we announce «... 883 million ...».

Last, $619^2 = 600 \cdot 638 + 19^2 = 382\,800 + 361 = 383\,161$. Add $404 + 383 = 787$; the rest of the answer is therefore «... 787 thousand 161». In figures: $258\,619^2 = 66\,883\,787\,161$.

The same technique can be extended to numbers of seven or more digits (yielding answers in the trillions and beyond), but the calculations become more challenging.

Chapter 15

Advanced multiplication

RETAINING MANY DIGITS in memory is the main obstacle to multiplying large numbers mentally. The mnemonic major system offers helpful support, and careful planning for efficiency can also relieve the burden on the memory. This chapter proves that multiplying large numbers in one's head is not only possible but also quite feasible, with good technique and sufficient practice.

Simplifying the factors

The various techniques for multiplying two-digit numbers can be applied or adapted to larger numbers. The following example uses both factoring and the difference of squares:

$$\begin{aligned} 126 \cdot 345 &= 63 \cdot 2 \cdot 345 \\ &= 63 \cdot 690 \\ &= (66-3)(66+3)10 \\ &= \left(66^2 - 3^2\right)10 \\ &= (4\,356 - 9)10 \\ &= 43\,470. \end{aligned}$$

Subtraction can be helpful:

$$796 \cdot 381 = (800 - 4)381$$

$$\begin{aligned}&= 800 \cdot 381 - 4 \cdot 381\\&= 304\,800 - 1\,524\\&= 303\,276.\end{aligned}$$

Subtraction, indeed, is just a special case of the principle that one may break a product into the sum of smaller products as one pleases. The previous chapter ended with $516 \cdot 619 = 500 \cdot 619 + (16 \cdot 600 + 16 \cdot 19)$. Multiplying separately by $500 = \frac{1}{2} \cdot 1\,000$ was easy, and the remaining multiplication by 16 could be done in various ways (others include $2 \cdot 8 \cdot 619$ and $4 \cdot 4 \cdot 619$).

Reüsing calculations

Any calculation that you have performed is yours to keep. Feel free to use it again. In the example above of $796 \cdot 321$, we could have computed $4 \cdot 381$ first and then doubled it to obtain $8 \cdot 381$, which also came up in the calculation.

You can calculate $3\,638 \cdot 53$ by first calculating $36 \cdot 53 = 1\,908$. Then

$$\begin{aligned}3\,638 \cdot 53 &= (36 \cdot 100 + 36 + 2)53\\&= [19, (08 + 19), 08] + 2 \cdot 53\\&= 192\,708 + 106\\&= 192\,814.\end{aligned}$$

Similarly,

$$\begin{aligned}3\,672 \cdot 53 &= (36 \cdot 100 + 36 \cdot 2)53\\&= [19, (08 + 38), 16]\\&= 194\,616.\end{aligned}$$

It pays to plan one's work instead of proceeding mechanically. Conventional arithmetic, performed on paper, uses a uniform method for all types of multiplication; mental arithmetic takes advantage of a variety of methods, some of which may be better for a given problem than others.

Cross-multiplication

The distributive property of multiplication with respect to addition enables us to break multiplication into pieces:

$$(a + b)(c + d) = ac + ad + bc + bd.$$

Indeed, the conventional algorithm used for multiplying by hand works by casting a product as the sum of numerous products of one-digit numbers.

The same principle enables us to multiply two three- or four-digit numbers by multiplying four one- or two-digit numbers. If a and b have two digits each, $100a + b$ is their four-digit concatenation (a followed by b). By the distributive property,

$$\begin{aligned}(100a + b)(100c + d) &= 10\,000ac + 100ad + 100bc + bd\\ &= 10\,000ac + 100(ad + bc) + bd.\end{aligned}$$

In simpler notation:

$$[a, b] \cdot [c, d] = [ac, (ad + bc), bd].$$

This method is known as *cross-multiplication* because its middle step produces a cross-shaped pattern, as is shown in the diagram below.[52]

52 Cross-multiplication is also a technique, equivalent to the conventional algorithm for manual multiplication, that proceeds from right to left with sums of the products of pairs of digits. It is a valid technique that can be applied easily to numbers of any length, but I do not consider it to be mental arithmetic if the answer is written down digit by digit during the calculation, as it typically is. It is described in many other sources; I do not present it here.

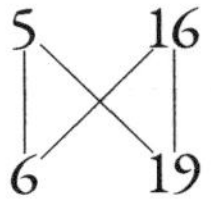

Using cross-multiplication for the previous example of $516 \cdot 619$, we proceed in this manner:

$$\begin{aligned}
516 \cdot 619 &= (500 + 16)(600 + 19) \\
&= 5 \cdot 6 \cdot 10\,000 + (5 \cdot 19 + 6 \cdot 16)100 + 16 \cdot 19 \\
&= 300\,000 + (95 + 96)100 + 304 \\
&= 300\,000 + 19\,100 + 304 \\
&= 319\,404.
\end{aligned}$$

Four-digit numbers can be multiplied in the same way. Here is the work in full; below is a description of a practical way to perform it mentally.[53]

$$\begin{aligned}
8\,723 \cdot 6\,694 &= 87 \cdot 66 \cdot 10\,000 + (87 \cdot 94 + 66 \cdot 23)100 \\
&\qquad + 23 \cdot 94 \\
&= 87 \cdot 6 \cdot 11 \cdot 10\,000 + (87 \cdot 94 + 66 \cdot 23)100 \\
&\qquad + 23 \cdot 94 \\
&= 5\,742 \cdot 10\,000 + \big(87(100 - 6) + 66 \cdot 23\big)100 \\
&\qquad + 23 \cdot 94 \\
&= 57\,420\,000 + \big((8\,700 - 522) + 66 \cdot 23\big)100 \\
&\qquad + 23 \cdot 94 \\
&= 57\,420\,000 + (8\,178 + 6 \cdot 253)100 + 23 \cdot 94 \\
&= 57\,420\,000 + (8\,178 + 1\,518)100 + 23 \cdot 94
\end{aligned}$$

53 This particular example would be better done with subtraction: $8\,723 \cdot 6\,694 = 8\,723(6\,700 - 6)$. The product $87 \cdot 67$ is $77^2 - 10^2 = 5\,829$, and already, seeing that there will be nothing to carry, we can announce «58 million ...». Cross-multiplication is used instead here as an illustration of a particularly laborious case; usually cross-multiplication is easier.

$$= 57\,420\,000 + 969\,600 + 23 \cdot 94$$
$$= 58\,389\,600 + 23(100 - 6)$$
$$= 58\,389\,600 + (2\,300 - 138)$$
$$= 58\,389\,600 + 2\,162$$
$$= 58\,391\,762.$$

Much of the bulk above consists of zeroes. The arithmetic is not so formidable as it may appear.

We start with $87 \cdot 66$, which we multiply as $87 \cdot 6 \cdot 11 = 522 \cdot 11 = 5\,742$. The answer is therefore 57 million or more, but we expect that there may be carrying into the millions' place, so we do not yet begin to announce the answer; instead, we preserver 5 742 as *elk horn*, *Leghorn*, *all green*, *walk! run!*, *liquor–wine*, or anything else that comes to mind.[54] Having encoded the number, we can now forget it and continue with the next step.

We multiply the two outer pairs of digits (87 and 94) and the two inner pairs (23 and 66). For $87 \cdot 94$, subtraction may be the easiest approach: $87(100 - 6) = 8\,700 - 522 = 8\,178$. We could, however, use 100 as a point of reference: $87 \cdot 94 = (100-13)(100-6) = 100(87-6)+13\cdot 6 = 8\,100+78 = 8\,178$. If necessary, we retain 8 178 as *foot-cuff* or *feed coffee*. For $23 \cdot 66$, factoring is again best: $23 \cdot 11 \cdot 6 = 253 \cdot 6 = 1\,518$. We add these products to get $8\,178+1\,518 = 9\,696$. The addition is most easily done two digits at a time: $[81+15, 78+18] = [96, 96] = 9\,696$.

The first step gave us 5 742 followed by four zeroes; the second step, 9 696 followed by two zeroes. We can add these numbers simply by adding the last two digits of 5 742 to the first two digits of 9 696: $[57, 42 + 96, 96] = [57, 42 + 100 - 4, 96] =$

54 My first idea was *licorne*, meaning 'unicorn' in French. Anything that you can remember will do. Do not waste time hunting for the ideal word or phrase; you will need it for only a moment, so take anything that will work and move on.

[58, 38, 96]. The last step of the calculation, $23 \cdot 94 < 2\,300$, will not cause carrying into the millions, so we now announce «58 million ...» as the first part of the answer and retain 3 896 in memory as *move a peach* or *mauve bush*.

The final step is $23 \cdot 94 = 23(100-6) = 2\,300-138 = 2\,162$. Again, we add the last two digits of 3 896 to the first two of 2 162 to obtain $[38, 96+21, 62] = [39, 17, 62]$, which gives us the rest of the answer: «... 391 thousand 762».

Condensed cross-multiplication

As is stated above, cross-multiplication of two four-digit numbers $[a, b]$ and $[c, d]$, in which each of a, b, c, and d has two digits, proceeds in the following manner:

$$[a, b] \cdot [c, d] = [ac, (ad + bc), bd].$$

A compact algorithm reduces the number of products from four to three.[55] Try it: you may like it.

Step 1: Compute ac and bd.

Step 2: Merge them as $[ac, bd]$. Perform all required carrying. The result will have three two-digit components, $[e, f, g]$. Remember these three numbers.

Step 3: Compute $h = f + g - (a - b)(c - d)$, which may be negative.

Step 4: The product is $[e, (e + f), h, g]$. Calculating it may require further merging of numbers. If h is negative, borrow just enough hundreds from $e + f$ to make h positive.

A few examples will illustrate the technique and the slight complications that can arise.

55 Ron Doerfler, «A Method for 4 × 4 Digit Mental Multiplications».

Previously we computed $8\,723 \cdot 6\,694 = 58\,391\,762$. Let us try the same thing with the simplified approach.

Step 1: We have $a = 87$, $b = 23$, $c = 66$, $d = 94$. Then $ac = 87 \cdot 66 = 29 \cdot 2 \cdot 3 \cdot 33 = 58 \cdot 99 = [58, -58] = 5\,742$, and $bd = 23 \cdot 94 = 23 \cdot 100 - 23 \cdot 6 = [23, -138] = [21, 62] = 2\,162$.

Step 2: We merge

$$[5742, 2162] = [57, 42 + 21, 62] = [57, 63, 62].$$

Remember these three figures; forget the rest. If necessary, create words for them, such as *lake, jam, chain*.

Step 3: First we compute $(87 - 23)(66 - 94) = 64(-28) = -64 \cdot 7 \cdot 4 = -448 \cdot 4 = -1\,792$. This must be subtracted from $63 + 62 = 125$. Thus we have $125 - (-1\,792) = 125 + 1\,792 = [18, 117] = 1\,917$.

Step 4: The product is

$$\begin{aligned}[57, (57 + 63), 1917, 62] &= [57, 120, 1917, 62] \\ &= [58, 39, 17, 62] \\ &= 58\,391\,762.\end{aligned}$$

That example was relatively difficult. More typical is $9\,277 \cdot 3\,478$.[56]

Step 1: Compute $92 \cdot 34 = 100 \cdot 34 - 8 \cdot 34 = [34, -272] = [31, 28] = 3\,128$ and $77 \cdot 78 = 77^2 + 77 = 5\,929 + 77 = [59, 106] = [60, 06] = 6\,006$.

Step 2: Merge:

$$[3128, 6006] = [31, 28 + 60, 06] = [31, 88, 06].$$

If necessary, these are *meat, fife, sash*.

Step 3: First, $(92 - 77)(34 - 78) = 15(-44) = 30(-22) = -660$. Then we have $88 + 6 - (-660) = 94 + 660 = 754$.

56 I produced this one and the next one with a random-number generator.

Step 4: The product is

$$\begin{aligned}[31, 31 + 88, 754, 06] &= [31, 119, 754, 06] \\ &= [32, 26, 54, 06] \\ &= 32\,265\,406.\end{aligned}$$

The number obtained at Step 3 is likely to be negative if either $a > b$ and $c > d$ or $a < b$ and $c < d$. That occurs in $6\,194 \cdot 3\,466$.

Step 1: We have $61 \cdot 34 = 60 \cdot 34 + 34 = 2\,040 + 34 = 2\,074$ and $94 \cdot 66 = 100 \cdot 66 - 6 \cdot 66 = [66, -396] = [62, 4] = 6\,204$.

Step 2: Merging yields $[2074, 6204] = [20, 74 + 62, 04] = [20, 136, 04] = [21, 36, 04]$. Remember as *nut, match, sewer*.

Step 3: First, $(61 - 94)(34 - 66) = (-33)(-32) = 32^2 + 32 = 1\,024 + 32 = 1\,056$ (or, using 30 as the point of reference, $30 \cdot 35 + 3 \cdot 2 = 1\,050 + 6 = 1\,056$). Second, $36 + 4 - 1\,056 = 40 - 1\,056 = -1\,016$.

Step 4: We have $[21, 21 + 36, -1\,016, 04]$. To eliminate the negative number, borrow the next higher number of hundreds, in this case 1 100:

$$\begin{aligned}[21, 57, -1\,016, 04] &= [21, 57 - 11, 1\,100 - 1\,016, 04] \\ &= [21, 46, 84, 04] \\ &= 21\,468\,404.\end{aligned}$$

This streamlined version of cross-multiplication does require practice. Although less natural and more contrived than the full method, it can be more efficient, and even quite rapid for some combinations of numbers. Consider adopting it. Note that it works as well if one or both numbers have only three digits.

Larger numbers

With numbers of five or more digits, cross-multiplication usually works best with the break at the three-digit point, which corresponds well to the English way of naming numbers. The example of $29\,832 \cdot 87\,576$ will demonstrate the technique.

First, we frame the problem as $(29\,000+832)(87\,000+576)$. The number of millions is $29 \cdot 87 = 30 \cdot 87 - 87 = 2\,610 - 87 = 2\,510 + 13 = 2\,523$. There will be no carrying of billions: the product is less than $30\,000 \cdot 90\,000 = 2.7$ billion, and in any event a two-digit number multiplied by a three-digit number yields less than 100 000, so the next step (with its two products) would add less than 200 000 to 523 000, which would not cause carrying. Accordingly, we now announce «2 billion ...» and remember 523 as *yellow gnome* or *wily enemy*.

Now we need $832 \cdot 87 + 576 \cdot 29$. If we are alert, we may notice that $87 = 29 \cdot 3$ and that $576 \cdot 29 = 192 \cdot 3 \cdot 29 = 192 \cdot 87$; therefore, $832 \cdot 87 + 576 \cdot 29 = (832 + 192)87 = 1\,024 \cdot 87 = 87\,000 + 24 \cdot 87 = 87\,000 + (2\,400 - 24 \cdot 13) = 87\,000 + 2\,088 = 89\,088$. Thus we can calculate the sum all at once, because the numbers happen to support a streamlined calculation.

If we do not notice that possibility, we compute the products separately and add the results. For $832 \cdot 87$, four approaches come to mind. If we know the multiplication table to 13, we can use subtraction: $832(100 - 13) = 83\,200 - [104, 39, 26] = 83\,200 - 10\,816 = 72\,384$. Almost as good, still using subtraction, is $832(90 - 3)$, for which we start by calculating $832 \cdot 3 = 2\,496$; then $832 \cdot 90 = 2\,496 \cdot 3 \cdot 10 = 74\,880$, and $74\,880 - 2\,496 = 74\,880 - 2\,500 + 4 = 72\,384$. The difference of squares can help: $832 \cdot 87 = 83 \cdot 87 \cdot 10 + 2 \cdot 87 = (85-2)(85+2)10+2 \cdot 87 = (7\,225-4)10+2 \cdot 87 = 72\,210+174 = 72\,384$. There is a hidden opportunity to use a point of reference: $832 \cdot 87 = \frac{1}{10}(832 \cdot 870) = \frac{1}{10}(800 \cdot 902 + 32 \cdot 70) = 80 \cdot 902 + 32 \cdot 7 = 72\,160 + 224 = 72\,384$. However we calculate it, we retain 72 384 as, say, *cane mover*.

As for $576 \cdot 29$, subtraction is probably best: $576(30 - 1) = 576 \cdot 30 - 576 = 17\,280 - 576 = 16\,704$. We then add in two steps: $[16 + 72, 704 + 384] = [88, 1\,088] = [89, 088] = 89\,088$.

The last step of the calculation, $832 \cdot 576 \approx 500\,000$, will not cause carrying of a million ($088 + 500 < 1\,000$), so we add 89 to the previous result of 523 (yielding 612) and announce «... 612 million ...», while retaining 088 as, say, *safe fee*.

Finally, we multiply $832 \cdot 576$. A well-chosen strategy can simplify this difficult task. We try factoring the smaller number: $576 = 8 \cdot 72 = 8 \cdot 8 \cdot 9$. Now we can multiply three times: $832 \cdot 9 = 7\,488$; $7\,488 \cdot 8 = 59\,904$; $59\,904 \cdot 8 = 479\,232$. Factoring the larger number also works well: $832 = 8 \cdot 104 \cdot 576 = 8(100 + 4)576 = 8 \cdot (57\,600 + 2\,304) = 8 \cdot 59\,904 = 479\,232$. We could also use the difference of squares by taking the average, $(832 + 576)/2 = 1\,408/2 = 704$, and representing $832 \cdot 576$ as $(704 + 128)(704 - 128) = 704^2 - 128^2 = 495\,616 - 16\,384 = 479\,232$. Cross-multiplication is another possibility: $832 \cdot 576 = 8 \cdot 5 \cdot 10\,000 + (76 \cdot 8 + 32 \cdot 5)100 + 32 \cdot 76 = 400\,000 + 76\,800 + 2\,432 = 479\,232$. All methods, of course, yield the same result.

It may be prudent to keep 232 in memory as *anemone* or *honeymoon* while we add $479 + 088 = 567$. We conclude by announcing «... 567 thousand 232». Written out in full, the product is 2 612 567 232.

Multiplication with so many digits is no task for the beginner: it requires facility with smaller products and sums, combined with the ability to retain numbers in memory with or without mnemonic assistance. With practice, however, a calculator can develop the ability to multiply large numbers mentally.

Chapter 16

Powers

INTEGRAL POWERS greater than 2 can sometimes be calculated exactly through the use of simple mental techniques. Powers yielding many digits and non-integral powers such as $x^{3.42}$ may have to be approximated; approaches to calculating them are given in the chapter on applications of logarithms.

Cubes

Often a small integer can be cubed most efficiently by multiplying twice. For instance, $6^3 = 6^2 \cdot 6 = 36 \cdot 6 = 216$, and $11^3 = 121 \cdot 11 = 1\,331$. The small cubes, however, occur so frequently that they should simply be learnt:

1	1	6	216
2	8	7	343
3	27	8	512
4	64	9	729
5	125	10	1 000

The cube of the general two-digit number $10a+b$, with digits a and b, is given by the following equation:

$$\begin{aligned}(10a+b)^3 &= (10a)^3 + 3(10a)^2 b + 3(10a)b^2 + b^3 \\ &= a^3 \cdot 1\,000 + (3ab)(10a+b) \cdot 10 + b^3.\end{aligned}$$

For example,

$$\begin{aligned} 37^3 = (10 \cdot 3 + 7)^3 &= 3^3 \cdot 1\,000 + (3 \cdot 3 \cdot 7) \cdot 37 \cdot 10 + 7^3 \\ &= 27\,000 + 21 \cdot 3 \cdot 37 \cdot 10 + 343 \\ &= 27\,000 + 21 \cdot 111 \cdot 10 + 343 \\ &= 27\,000 + 23\,310 + 343 \\ &= 50\,653. \end{aligned}$$

Another equation:

$$(10a + b)^3 = (10a)(10a + b)(10a + 2b) + (10a + b)b^2.$$

This equation resembles the use of a point of reference for multiplication. To compute 37^3, we note that $37 + 3 = 40$ and therefore also use $37 - 3 = 34$:

$$\begin{aligned} 37^3 &= 40 \cdot 37 \cdot 34 + 37 \cdot 3^2 \\ &= 1\,258 \cdot 40 + 111 \cdot 3 \\ &= 50\,320 + 333 \\ &= 50\,653. \end{aligned}$$

Another example, using a larger number:

$$\begin{aligned} 94^3 &= 90 \cdot 94 \cdot 98 + 94 \cdot 4^2 \\ &= 90(96^2 - 2^2) + 94 \cdot 16 \\ &= 90(9\,212) + 94 \cdot 16 \\ &= 829\,080 + 1\,504 \\ &= 830\,584. \end{aligned}$$

Note from this example that cubes of even numbers lend themselves to the use of the difference of squares.

Fourth powers

A fourth power is a square of a square. The fourth powers of the single digits all come from memory, if one knows the squares up to 100: $9^4 = (9^2)^2 = 81^2 = 6\,561$.

The fourth power of a two-digit number is the square of a three- or four-digit number:

$$\begin{aligned} 87^4 &= (87^2)^2 \\ &= 7\,569^2 \\ &= 8\,000 \cdot 7\,138 + 431^2 \\ &= 57\,104\,000 + 400 \cdot 462 + 31^2 \\ &= 57\,104\,000 + 184\,800 + 961 \\ &= 57\,289\,761. \end{aligned}$$

Fourth powers of three-digit numbers are feasible:

$$\begin{aligned} 139^4 &= (139^2)^2 \\ &= (17\,800 + 1\,521)^2 \\ &= 19\,321^2 \\ &= 19^2 \cdot 10^6 + 2 \cdot 19 \cdot 321 \cdot 10^3 + 321^2 \\ &= 361 \cdot 10^6 + 2 \cdot 6\,099 \cdot 10^3 + 321^2 \\ &= 361\,000\,000 + 12\,198\,000 + 300 \cdot 342 + 21^2 \\ &= 373\,198\,000 + 102\,600 + 441 \\ &= 373\,301\,041. \end{aligned}$$

Reciting the answer orally, we would see that nothing would be carried into the millions, so we would state «373 million» as soon as we had calculated the second term of the five-digit square. If necessary, we could hold 198 in memory as *top off* or *hot beef*, perhaps, while calculating 341^2; then the thousands would be $198 + 102 = 300$, plus 1 carried from $600 + 441$.

Higher powers of small numbers

Calculating 2^{20} by multiplying 20 factors of 2 would be tedious and prone to error. Powers of powers can simplify the work:

$$\begin{aligned}
2^{20} &= 2^{10\cdot 2} \\
&= (2^{10})^2 \\
&= (2^{5\cdot 2})^2 \\
&= \left((2^5)^2\right)^2 \\
&= \left((2^3 \cdot 2^2)^2\right)^2 \\
&= \left((8 \cdot 4)^2\right)^2 \\
&= (32^2)^2 \\
&= 1\,024^2 \\
&= 1\,048\,576.
\end{aligned}$$

To calculate 3^{12}, note that it is equal to $(3^6)^2$. For 3^6, calculate either $3^6 = (3^3)^2 = 27^2 = 729$ or $3^6 = (3^2)^3 = 9^3 = 9 \cdot 9 \cdot 9 = 81 \cdot 9 = 729$. Complete the calculation with $729^2 = 758 \cdot 700 + 29^2 = 530\,600 + 841 = 531\,441$.

Finally, try 7^5:

$$\begin{aligned}
7^5 &= 7^{1+4} \\
&= 7(7^4) \\
&= 7\left((7^2)^2\right) \\
&= 7(49^2) \\
&= 7(2\,401) \\
&= 16\,807.
\end{aligned}$$

Within practical limits, these techniques can be extended to two-digit numbers. Higher powers of numbers that are too large for exact mental calculation can still be approximated with logarithms; a good method appears later in this book.

Chapter 17

Roots

GRINDING A SQUARE ROOT OUT digit by digit in the traditional way is laborious on paper and impractical for mental calculation. Although better algorithms for calculating one digit at a time exist, and some are suitable for mental work, a different technique can quickly yield six digits or more with less effort. The indefatigable mental arithmetician will also find in this chapter some good ways to extract cube roots and roots of even higher degree.

Square roots

The method presented here normally yields at least six correct digits—more than enough for most practical purposes. (Sometimes the sixth digit is slightly inaccurate.) To extract the square root of a number R, we obtain an estimate x and refine it by adding the approximate error, which, by the Newton–Raphson method,[57] is

$$\frac{R - x^2}{2x}.$$

57 As applied to square roots, this method is equivalent to the use of averages that is presented in the chapter on estimation. To see square roots calculated to many digits through clever use of averages, read AC Aitken's article «The Art of Mental Calculation: With demonstrations», which is cited in the bibliography.

This refinement approximately doubles the number of correct digits. If our estimate has three correct digits, the refined estimate will probably have six.

We begin by making a note of the position of the decimal point in the answer. If the number of digits to the left of the decimal point is even, there will be half as many digits in the square root; if odd, there will be one more than half as many, disregarding the remainder.[58] Thus there are 3 digits in the square root of any 5- or 6-digit number. If the radicand is less than 1, there will be one zero after the decimal point for each pair of zeroes in the radicand; for example, the square root of 0.00 00 04 will start with 0.00 (it is in fact 0.002).

For most of the rest of the calculation, we shall work in integers; only at the end shall we insert the decimal point.

Most methods for calculating a square root start with determination of the first digit. Using our knowledge of the squares up to 100, we mental calculators can immediately give the first *two* digits of the root. We take the first three digits of the radicand if the number of digits to the left of the decimal point is odd, the first four digits if it is even. (If it is less than 1, we take the first pair of digits that are not both zeroes, together with the pair following it.) The first two digits of the root are the highest number whose square is less than or equal to those digits.

These first two digits give us our initial estimate. We obtain a third digit by subtracting the square of our estimate from the first group of digits and dividing by twice the estimate. Since we need only more digit at this point, the division can be performed coarsely.

Having three digits, we get three more by repeating the process: subtract the square of the estimate from the radicand and divide by twice the estimate. If the estimate is close to the correct

58 More simply put, the number of digits in the answer is half of the number of digits in the integral part of the radicand, rounded up.

root (especially if the fourth digit that we obtain is a 0), we may get *more than* six correct digits.

A few examples will demonstrate the method.

Suppose that we wish to calculate $\sqrt{123\,456}$. It is a 6-digit number; therefore, the root has $6 \div 2 = 3$ digits. We take the first four digits, 1234, because the number of digits is even. What is $\sqrt{1\,234}$? From memory, $35^2 = 1\,225$, and $36^2 = 1\,296$; therefore, the first two digits are 35, and the root is closer to 350 than to 360.

For the third digit, we calculate $1\,234 - 1\,225 = 9$ and append the next pair of digits, 56. This figure, 956, we divide by twice our estimate: $2 \cdot 350 = 700$. Since we need only one-digit accuracy, we get the digit 1 and stop. Our new estimate is 351, and we can now call off «351 point ...» as the first part of the answer. Anyone listening will be impressed at our obtaining three digits so quickly.

Now, in principle, we need to calculate

$$\frac{123\,456 - 351^2}{2 \cdot 351}.$$

It is easier, however, to make use of the work already done. After all, we have calculated 350^2 and subtracted it from 123 456. Instead of computing 351^2 from scratch, we can take advantage of the fact that

$$(a + b)^2 = a^2 + 2ab + b^2,$$

which implies that

$$351^2 = (350 + 1)^2 = 350^2 + 2 \cdot 350 \cdot 1 + 1^2.$$

Consequently, from the remainder 956 already obtained, we need only subtract $2 \cdot 350 \cdot 1 + 1^2$. Rather than calculating this sum, we subtract each part in turn: $956 - 700 = 256$, then $256 - 1 = 255$. Proceeding in this way lightens the burden on our memory.

To obtain three more digits of the root, we divide this difference of 255 by twice the current root, namely $2 \cdot 351 = 702$. To avoid worrying about the decimal point, we can replace 255 with 255 000 for our three-digit division. The three zeroes turn our refinement from a decimal into an integer. If there were more digits in the radicand, we would bring the next three down instead of using zeroes.

Since $702 \approx 700 + \frac{1}{3}\%$, we can approximate $255\,000/702$ as $255\,000/700 - \frac{1}{3}\%$. To three places, $255\,000/700 = 364$; then 1% of that figure is about 4 (rounded), so $\frac{1}{3}\%$ is about 1. We thus obtain $364 - 1 = 363$ as the last three digits of our refined estimate, which is therefore 351363.

Now is the time to place the decimal point—but we have already done this: 351.363. We call out «3 6 3» as the last digits of our answer, which turns out to be correct to six figures.

If instead we wish to calculate $\sqrt{12\,345}$, we note that it, too, has 3 digits: 1 more than $5 \div 2$ without the remainder. We take the first three digits of the radicand, 123. Their square root is 11, because $11^2 = 121$ and $12^2 = 144$. Already we can tell that the root is slightly greater than 110.

How much greater? To obtain the next digit, we compute $123 - 121 = 2$ followed by the next two digits of the radicand, 45. Then 245 divided by $2 \cdot 110 = 220$ gives 1 as the next digit. Our estimate is therefore 111, and since we know that the root contains 3 digits, we now call out the first part of the answer, «111 point ...».

Again we note that $111^2 = (110+1)^2 = 110^2 + 2 \cdot 110 \cdot 1 + 1^2$. We have already subtracted 110^2; now we must subtract what remains, 221, from 245, and append three zeroes. We then divide the result, 24 000, by $2 \cdot 111 = 222$. It may help to cancel a factor of 2 to obtain 12 000/111. This is easily computed with long division: 120/111 is 1 remainder 9; 90/111 is 0 remainder 90; 900/111 is approximately 8 (we need not carry the calculation beyond this our third digit, which should be rounded appropriately). Our three additional digits are therefore 108, which we now call out; the root is 111.108, correct to six figures.

❧

For our final example, let us approximate $\sqrt{\pi}$ by taking π to be 3.141 592 7.[59] The answer will have a single digit before the decimal point. For $\sqrt{314}$, we get 17, whose square is 289. Subtract: $314 - 289 = 25$. Bring two more digits down to get 2 515. This divided by $170 \cdot 2 = 340$ is 7. We can call out «1.77...».

From 2 515, we subtract $2 \cdot 170 \cdot 7 = 2\,380$, obtaining 135, and then subtract $7^2 = 49$, obtaining 86. Since there are more digits in the radicand, we bring them down rather than appending three zeroes. The last three digits of the root will come from 86 927 divided by $177 \cdot 2 = 354$. The 5 in the middle of 354 encourages us to double dividend and divisor in the expectation of simplifying the division. We obtain 173 854/708, which we approximate as 173 900/700 − 1%. (Since we shall end the calculation at three digits, we can round the numerator off in this way.) This is 248 − 1% ≈ 246. We call those three digits out and conclude that the root is 1.772 46 to six figures.

Unfortunately, inaccuracy crept into the last digit: the root is actually 1.772 45. The sloppy division at the end might be suspected of causing the slight error, but even 173 854/708 would have been rounded to 246. A small error in the last digit does occur sometimes, most often because of rounding. In this example, the division at the end yields $245.\underline{5}56\ldots$; the root continues $\ldots 245\underline{3}86$. The inaccuracy, therefore, was in the *seventh* digit

9 The last digit is rounded up: $\pi \approx 3.141\,592\,65\ldots$.

(underlined), which caused rounding up in our calculation but down in the actual root.

It is not unusual for the last digit of an estimate to be slightly off. That is in the nature of an estimate. If you are called out for an «error», explain that you calculate six digits quickly and cannot guarantee the correctness of the last one. You could instead calculate eight digits, by refining a four-digit estimate; the first seven would almost always be correct, and the eighth would be wrong sometimes (though at other times you would get nine or ten correct digits). That, however, would require considerable extra work for accuracy that would seldom be needed. Already five or six correct figures exceed the usual needs of engineers.

Other roots

One day, to make a point at work, I told someone that a certain task was so easy that I could perform it while calculating a cube root in my head, without anyone's noticing that my attentions were elsewhere. I arbitrarily wrote $\sqrt[3]{123\,456}$ on a scrap of paper and set about fulfilling my promise.

Immediately I noticed that the number happened to be quite convenient: $125 = 5^3$, so $125\,000 = 50^3$, and $\sqrt[3]{123\,456}$ was just under 50. I wrote «49.» (including the decimal point) and began to seek more digits.

Using differentials from the differential calculus, it is easy to show that $\sqrt[n]{x+\delta}$, when δ is small relative to x, can be estimated well from a known root $r = \sqrt[n]{x}$ as follows:

$$\sqrt[n]{x+\delta} \approx r + \frac{\delta}{nr^{n-1}}.$$

In other words, we correct for δ by adding an adjustment to r.

For the calculation at hand, $n = 3$, $x = 125\,000$, $\delta = -1\,544$ (*i.e.*, $123\,456 - 125\,000$), and $r = 50$. We have

$$\begin{aligned}\sqrt[3]{123\,456} &\approx 50 + \frac{-1\,544}{3 \cdot 50^2}\\ &\approx 50 - \frac{515}{2\,500}\\ &= 50 - \frac{2060}{10\,000}\\ &= 50 - 0.206\\ &= 49.794.\end{aligned}$$

The root to five decimal places is actually 49.793.

I simplified this result to 49.8, for convenient calculation, and repeated the process. By the binomial theorem,[60]

$$(a-b)^3 = a^3 - 3a^2b + 3ab^2 - b^3.$$

Therefore,

$$\begin{aligned}49.8^3 &= (50 - 0.2)^3\\ &= 125\,000 - 3 \cdot 2500 \cdot 0.2 + 3 \cdot 50 \cdot 0.04 - 0.008\\ &= 123\,505.992\\ &\approx 123\,506.\end{aligned}$$

This cube exceeds the radicand by 50 ($123\,456 - 123\,506$). I therefore obtained as the new approximation of the root $49.8 + \frac{-50}{3 \cdot 49.8^2}$. Because $498^2 = 248\,004$ (using the approach for a square near 500), I found that $49.8^2 = 2\,480.04 \approx 2\,480$. The new root was therefore approximately

$$49.8 + \frac{-50}{3 \cdot 49.8^2} \approx 49.8 - \frac{50}{3 \cdot 2\,480}$$

60 The binomial theorem is explained in an appendix.

$$= 49.8 - \frac{50}{7\,440}$$
$$\approx 49.8 - 0.006\,720$$
$$= 49.793\,280.$$

I used long division for $\frac{50}{7\,440}$ (after simplifying it to $\frac{5}{744}$), since I needed as many correct digits as possible and certainly did not want to spoil the calculation right at the end. Had an approximation sufficed, I could have taken $\frac{5}{750}$ + 0.8%; it happens to be 0.006 720, the actual quotient obtained through long division.

To nine digits, the actual root is 49.793 279 8; rounding to eight digits gives the number that I obtained. The cube root that I calculated mentally while performing a different task at work was therefore accurate to 4 parts in a billion.

Lest you suppose that this calculation bears witness to great skill, I freely admit that I was lucky: the root happened to be close to 50, a convenient number for calculation (mainly because doubling it yields 100). Most radicands would have required messier calculations that I might not have continued for long.

The late but celebrated physicist Richard Feynman had the same good fortune when someone wielding an abacus challenged him to calculate $\sqrt[3]{1\,729.03}$. From experience with great grosses, cubic feet and cubic inches, and the like, he knew that 1 728 was 12^3. Using differentials, he found mentally that

$$\sqrt[3]{1\,729.03} \approx 12 + \frac{1.03}{3 \cdot 12^2} \approx 12 + \frac{1}{432} \approx 12.002.$$

He had written down half a dozen digits or so before his challenger laboriously clattered out three on the abacus.[61] With most other radicands, neither he nor I would have fared nearly so well.

61 Richard P Feynman, *"Surely You're Joking, Mr. Feynman!": Adventures of a curious character* (New York: W W Norton, 1985) at 196–98. I too would

I would not ordinarily calculate a cube root in my head to so many figures; I would use logarithms instead. The techniques appear in the chapter on computational applications of logarithms.

A fourth root can be computed as the square root of the square root. This may be a practical approach if you shrink from using logarithms, or if you need only a few digits' accuracy.

Roots of fifth or higher degree can be calculated in a manner akin to the one illustrated above for the cube root, but again logarithms are more practical even though they, too, can be cumbersome for mental calculation.

If we are satisfied with only a few digits' accuracy, however, we can use the aforementioned formula

$$\sqrt[n]{x + \delta} \approx r + \frac{\delta}{nr^{n-1}},$$

in which $r = \sqrt[n]{x}$ is a known root and δ is small relative to x.[62] An example will demonstrate this method of approximation.

To approximate $\sqrt[7]{2\,000}$, we first need a reasonable estimate to one or two digits. Clearly 10 is much too large, because 10^7 greatly exceeds 2 000; but 2 is too small, because 2^7 is only 128. We test 3 by computing $3^7 = 3 \cdot 3^6 = 3(3^3)^2 = 3(27^2) = 3 \cdot 729 = 2\,187$: although too high (because $3^7 > 2\,000$), 3 is

have recognised that the radicand was only slightly greater than 12^3; I also would have known that 1 729 is the smallest positive integer that can be represented as a sum of cubes of positive integers in two ways, namely $12^3 + 1^3$ and $10^3 + 9^3$, of which the former is useful for estimating the cube root.

62 If you know the differential calculus, you may benefit from this explanation. The derivative $f'(x)$ of a function $f(x)$ is defined as $\lim_{\Delta x \to 0}\big(f(x + \Delta x) - f(x)\big)/\Delta x$, if the limit exists. When Δx is sufficiently small, therefore, $f'(x) \approx \big(f(x + \Delta x) - f(x)\big)/\Delta x$. Hence $f(x + \Delta x) \approx f(x) + f'(x)\Delta x$. For the nth root, we have $f(x) = x^{1/n}$; therefore, $f'(x) = \frac{1}{n}x^{(1/n)-1} = \frac{1}{n}x^{-(n-1)/n} = \frac{1}{n}(x^{-1/n})^{n-1}$. Taking $\delta = \Delta x$ and $r = \sqrt[n]{x}$, we obtain $\sqrt[n]{x + \delta} \approx r + \delta/nr^{n-1}$.

good enough as the preliminary estimate of the seventh root. If giving the answer aloud, we can now say «2 point ...», because we know that the root is slightly less than 3.

For the next step, we apply the formula with $n = 7$ (the degree of the root), $r = \sqrt[7]{2\,187} = 3$, and $\delta = 2000 - 2187 = -187$. Two more accurate digits are the most that we can expect from this estimate, so the calculations can be conducted somewhat roughly.

$$\begin{aligned}
\sqrt[7]{2\,000} &\approx 3 + \frac{-187}{7(3^6)} \\
&= 3 - \frac{187}{7 \cdot 729} \\
&= 3 - \frac{187}{5\,103} \\
&\approx 3 - \frac{187 - 2\%}{5\,103 - 2\%} \\
&= 3 - \frac{183}{5\,000} \\
&= 3 - \frac{366}{10\,000} \\
&= 3 - 0.037 \\
&= 2.963.
\end{aligned}$$

The root is 2.9619..., so our estimate is accurate to two decimal places, and three digits overall. It is not off by much in the fourth digit.

We are not always so fortunate as to stumble upon a good approximation. More difficult is $\sqrt[5]{5\,000}$. To estimate the root, we note that 10^5 is 20 times the radicand; therefore $10^5/32 = 10^5/2^5 = (10/2)^5 = 5^5$ is less than the radicand, and the root lies between 5 and 10. We calculate $5^5 = 5(25^2) = 5 \cdot 625 = 3\,125$, which is too far from 5 000 for 5 to be a good estimate. We try instead $6^5 = 6(36^2) = 6 \cdot 1\,296 > 5\,000$: there is no need to finish the calculation, because $6 \cdot 1\,296 \approx 6 \cdot 1\,300 = 7\,800$ is obviously too great. From these calculations, however, we see that the root is roughly halfway between 5 and 6, so we try 5.5. Raising that two-digit number to the fifth power seems like a burdensome task, so we look for a better way and find one:

$$\begin{aligned}
5.5^5 &= \left(\frac{11}{2}\right)^5 \\
&= \frac{11^5}{2^5} \\
&= \frac{(11 \cdot 121^2)}{8 \cdot 4} \\
&= \frac{11 \cdot 14\,641}{8 \cdot 4} \\
&= \frac{[1, 5, 10, 10, 5, 1]}{8 \cdot 4} \\
&= \frac{161\,051}{8 \cdot 4} \\
&\approx \frac{20\,131}{4} \\
&\approx 5\,033.
\end{aligned}$$

Since our method will probably yield only three digits' accuracy, we can truncate the calculation at four digits or so; thus we stop at the decimal point when dividing by 8 and 4.

Our estimate is good: it is in excess by only 33, which is less than 1%. (Had it been poor, we would have selected a better one in proportion to the distance between 5 000 and either 5^5 or 6^5, whichever was nearer.) To apply the formula, where we need to divide by 5.5^4, it is easier to multiply numerator and denominator by 5.5 so that we can divide instead by the known value $5.5^5 \approx 5\,033$. This approach of multiplying numerator and denominator by the estimated root should be used generally, except where the denominator is easily calculated.

$$\begin{aligned}
\sqrt[5]{5\,000} &\approx 5.5 + \frac{-33}{5(5.5^4)} \\
&= 5.5 - \frac{33 \cdot 5.5}{5(5.5^5)} \\
&= 5.5 - \frac{33 \cdot 11}{10 \cdot 5\,033} \\
&= 5.5 - \frac{36.3}{5\,033} \\
&\approx 5.5 - \frac{36.3 - 0.7\%}{5\,033 - 0.7\%} \\
&= 5.5 - \frac{36}{5\,000} \\
&= 5.5 - \frac{72}{10\,000} \\
&\approx 5.5 - 0.007 \\
&= 5.493.
\end{aligned}$$

The root is 5.4928…, so all four digits of our estimate are correct. Indeed, had we subtracted 0.0072 (as the calculation suggested) rather than rounding it at the third decimal place, we would have had five correct digits; but this would have been the fruit of fortunate rounding and of the choice of a particularly good initial estimate. In general, a good but not great initial estimate should yield three correct digits.

❧

Now you know how to calculate higher roots directly. To put this knowledge into practice, you will need to develop your skill, not to mention your stamina. If you do so, you will possess a very impressive ability to do something that few mental calculators ever attempt.

What some mental calculators attempt, to great fanfare, is a «calculation» such as the 13th root of a 100-digit number. This they do in only a few minutes, because it is a gimmick: the root in question is always an integer, and that fact, thanks to a bit of number theory, gives an easy way to determine the first digit or two and the last few digits as well. The other digits can be worked out with logarithms or other expedients. The person performing the «calculation» examines only the first and last few digits of the radicand. In real life, we would hardly ever know *ab initio* that the root would come out even; we would just set about calculating as many digits as were needed.

Chapter 18

Logarithms

COMPUTATION of logarithms with paper and pencil is seldom taught or done. With practice, however, you can learn to compute logarithms to five decimal places entirely in your head. You may benefit from practising on paper before attempting mental calculation.

Unless you are confident of understanding logarithms, you may wish to review the section on logarithms in the first chapter before continuing.

Important logarithms to learn

Memorise the logarithms of as many small prime numbers as you wish, but at least those of 2, 3, and 7. The following logarithms occur frequently:

log 2	0.30103
log 3	0.47712
log 5	0.69897
log 7	0.84510
log 11	1.04139
log 13	1.11394
log 17	1.23045
log 19	1.27875
log 23	1.36173

You may learn only the mantissa, since the characteristic is obvious. Also, you may omit log 5, which can be calculated from log 2: $\log 5 = \log \frac{10}{2} = \log 10 - \log 2 = 1 - \log 2 = 1 - 0.30103 = 0.69897$. In practice, it is usually easier to subtract 0.30103 than to add 0.69897, so log 2 is used more frequently than log 5 in mental calculation.

The following logarithms are important for conversions between natural and common logarithms, which arise in mental calculation because sometimes we calculate a natural logarithm first:

ln 10	2.3023 (actually 2.3026)
log *e*	0.4343 (actually 0.43429)

For convenience in calculation, I have substituted $2.3023 = 23 \cdot 0.1001$ for ln 10 and $0.4343 = 43 \cdot 0.0101$ for log *e*. As will be seen, the slight error in the last digit usually does not affect the calculation of mantissas to five decimal places, because typically these two logarithms are multiplied by a number less than 1 that pushes the error beyond the fifth place.

Using prime factors

Because $\log ab = \log a + \log b$, the logarithm of an integer is the sum of the logarithms of its prime factors. The more logarithms of primes we know, the more logarithms of composite numbers we can compute by simple addition. Examples:

$$\begin{aligned}\log 42 &= \log 2 \cdot 3 \cdot 7\\ &= \log 2 + \log 3 + \log 7\\ &= 0.30103 + 0.47712 + 0.84510\\ &= 1.62325.\end{aligned}$$

Apply the rule $\log a^b = b \log a$ for a repeated factor:

$$\begin{aligned}\log 8 &= \log 2^3 \\ &= 3 \log 2 \\ &= 3 \cdot 0.30103 \\ &= 0.90309.\end{aligned}$$

The following example shows the substitution of $1 - \log 2$ for $\log 5$. Subtracting 0.30103 is easier than adding 0.69897. In a calculation such as this, one need only compute the mantissa; the characteristic (1, because $\log 55$ is between $\log 10 = 1$ and $\log 100 = 2$) can easily be added at the end. Thus we could have calculated $\log 11 - \log 2$ instead and changed the characteristic to 1.

$$\begin{aligned}\log 55 &= \log 5 \cdot 11 \\ &= \log 5 + \log 11 \\ &= (1 - \log 2) + \log 11 \\ &= 2.04139 - 0.30103 \\ &= 1.74036.\end{aligned}$$

Similarly, $\log 550 = 2.74036$, and $\log 0.055 = \bar{2}.74036 = 0.74036 - 2 = -1.25964$. It is easiest to carry only the mantissa in one's head and let its digits turn over like the numbers on a car's odometer, without a thought for the characteristic:

$$\begin{aligned}\log 55 &= \log 5 + \log 11 \\ &\to \log 11 - \log 2 \\ &\to 04139 - 30103 \\ &\to 74036 \\ &\to 1.74036.\end{aligned}$$

The arrow appears in place of the equality sign in statements that are mathematically incorrect but chosen for practical and efficient mental arithmetic. By thinking in this minimal way, we clear away details that we can easily address later, and we reduce the burden on our memory. Calculation of log 550 or log 0.055 proceeds in the same way: $\log 550 \rightarrow \log 5 + \log 11 \ldots.$

Because $\log \frac{a}{b} = \log a - \log b$, we can use primes to calculate the logarithms of many fractions:

$$
\begin{aligned}
\log \frac{2}{3} &= \log 2 - \log 3 \\
&\rightarrow 30103 - 47712 \\
&\rightarrow 82391 \\
&\rightarrow \bar{1}.82391 \\
&\rightarrow -0.17609.
\end{aligned}
$$

Remember that $\bar{1}.82391 = -1 + 0.82391$. If the logarithm is to be used in further calculations, it can be left in the form $\bar{1}.82391$; otherwise, it should be converted to −0.17609.

Five-digit addition and subtraction can tax the memory, but with memorised values they are less difficult than they may seem. Add the first two digits and then the last three, or the first three and then the last two. Remember to subtract $\log 2 = 0.30103$ rather than adding $\log 5 = 0.69897$. For log 13, add 11400 and subtract 6. With practice, you will be able to add the logarithms of primes in your head.

Interpolation

The *natural* logarithm of a number close to 1 can be calculated with this equation, which is valid only when $-1 < x \leq 1$:

$$\ln(1 + x) = x - \frac{x^2}{2} + \frac{x^3}{3} - \frac{x^4}{4} + \cdots.$$

Because x is small and we are calculating logarithms to only five decimal places, we can almost always simplify this expression (when $|x| < 0.08$) to

$$\ln(1+x) \approx x - \frac{x^2}{2} + \frac{x^3}{3}$$

and usually (when $|x| < 0.03$) to

$$\ln(1+x) \approx x - \frac{x^2}{2}.$$

When $|x| < 0.01$, the calculation becomes easy indeed:

$$\ln(1+x) \approx x.$$

For example,

$$\begin{aligned}\ln 1.06 &\approx 0.06 - \frac{0.06^2}{2} + \frac{0.06^3}{3} \\ &= 0.06 - \frac{0.0036}{2} + \frac{0.00020}{3} \\ &\approx 0.06 - 0.0018 + 0.00007 \\ &= 0.05827.\end{aligned}$$

Notice that the last term (0.00007) can be estimated quite roughly, since we are computing logarithms to only five digits in the mantissa. Although the example above coarsely estimates 0.06^3 as 0.00020 rather than giving the exact value 0.000216, division by 3 still yielded 0.00007, correctly, to five decimal places.

The *common* logarithm of a number close to 1 can be calculated by converting the natural logarithm:

$$\log(1+x) = (\log e)\ln(1+x) \approx 0.4343\left(x - \frac{x^2}{2} + \frac{x^3}{3}\right).$$

Again, if x is sufficiently small, we can drop some of the terms in parentheses. Note that the slight inaccuracy in 0.4343 (the correct value, to five digits, is 0.43429) is lost because multiplication by x, a small number, causes it to fall beyond the fifth decimal place.

For example,

$$\begin{aligned}\log 1.02 &\approx 0.4343\left(0.02-\frac{0.02^2}{2}\right)\\&=0.4343\left(0.02-\frac{0.0004}{2}\right)\\&=0.00869-0.00009\\&=0.00860.\end{aligned}$$

The common logarithm of 1.01 occurs frequently in the mental computation of logarithms and antilogarithms; it should be learnt.

$$\begin{aligned}\log 1.01 &\approx 0.4343\left(0.01-\frac{0.01^2}{2}\right)\\&=0.4343\left(0.01-\frac{0.0001}{2}\right)\\&=0.00434-0.00002\\&=0.00432.\end{aligned}$$

It is very close to $0.4343 \cdot 0.01$.

Logarithms of numbers slightly greater than 1 prove convenient in calculations such as this, which is correct as given:

$$\begin{aligned}\log 59 &= \log(60-1)\\&=\log 60\left(1-\frac{1}{60}\right)\\&=\log 60\left(1+\frac{-1}{60}\right)\\&\approx \log 60+0.4343\left(\frac{-1}{60}-\frac{(-1/60)^2}{2}\right)\end{aligned}$$

$$
\begin{aligned}
&= 1.77815 - \frac{0.4343}{60} - \frac{0.4343}{60 \cdot 60 \cdot 2} \\
&= 1.77815 - 0.00724 - 0.00006 \\
&= 1.77085.
\end{aligned}
$$

The term $(-0.4343(-1/60)^2)/2$ is most easily calculated by dividing the previous term $(0.4343/60)$ by $60 \cdot 2$. The result is a single digit in the fifth decimal place. It can be calculated coarsely, because only the first digit is needed. Note also the careful management of the negative signs.

Another approximation

Another good approximation, where δ (delta) is small relative to x:

$$\ln(x + \delta) \approx \ln x + \frac{2\delta}{2x + \delta} = \ln x + \frac{\delta}{x + \frac{\delta}{2}}.$$

The two expressions on the right are equivalent. One of them may be more convenient for a given problem.

In the following calculation of log 41, with $x = 40$ and $\delta = 1$, the first expression equals $2/81$; the second, $1/40\frac{1}{2}$. Although these expressions are equivalent, the first is more convenient for calculation. The answer is correct to five decimal places.

$$
\begin{aligned}
\log 41 &= \log(40 + 1) \\
&= \log 40 \left(1 + \frac{1}{40}\right) \\
&= \log 40 + \log \left(1 + \frac{1}{40}\right) \\
&\longrightarrow 1.60206 + (\log e) \ln \left(1 + \frac{1}{40}\right) \\
&\longrightarrow 1.60206 + 0.4343 \cdot \frac{2 \cdot 1}{2 \cdot 40 + 1}
\end{aligned}
$$

$$\begin{aligned}
&\to 1.60206 + 0.4343 \cdot \frac{2}{81} \\
&\to 1.60206 + \frac{0.8686}{81} \\
&\to 1.60206 + 0.01072 \\
&= 1.61278.
\end{aligned}$$

The calculation is shown in full for instructional purposes. In practice, one would begin by calculating $\log(1 + \frac{1}{40})$; the answer, 0.01072, could be remembered quite conveniently as «ten seventy-two». Then would come the easy calculation of $\log 40 = \log(2^2 \cdot 10) = 2\log 2 + \log 10 = 2 \cdot 0.30103 + 1 = 1.60206$, followed by addition: 1.6, 02 + 10, 06 + 72. Piecemeal calculation in this manner reduces the burden on the memory.

Suppose that we needed log 13 and did not know its value. Neither log 12 nor log 14 would be close enough for a good approximation. We could consider instead log 130, which differs from log 13 only in the characteristic; the mantissa is the same. A number especially near 130 is $128 = 2^7$. We would proceed in this manner, leaving the characteristic for the end as usual:

$$\begin{aligned}
\log 130 &= \log(128 + 2) \\
&= \log 128 \left(1 + \frac{2}{128}\right) \\
&= \log 2^7 + \log\left(1 + \frac{1}{64}\right) \\
&= 7\log 2 + \log\left(1 + \frac{1}{64}\right) \\
&\to 10721 + 43430 \ln\left(1 + \frac{1}{64}\right) \\
&\to 10721 + 43430 \cdot \frac{2 \cdot 1}{2 \cdot 64 + 1} \\
&\to 10721 + 43430 \cdot \frac{2}{128 + 1}
\end{aligned}$$

$$\begin{aligned}&\rightarrow 10721 + \frac{86860}{129}\\&\rightarrow 10721 + 00673\\&\rightarrow 11394\\&\rightarrow 2.11394.\end{aligned}$$

Therefore $\log 13 = 1.11394$. As you may know from memory, all digits of this answer are correct.

Note that the common logarithms of integral powers of 2 are easy to calculate, thanks to the zeroes and the other small digits in $\log 2 = 0.30103$.[63] Even $\log 2^{33} = 33 \log 2 = 9.93399$ can be calculated without carrying anything. For this reason, it is best to do the difficult part of the above calculation first, obtaining 673 (which is easy to remember: if necessary, convert it to *choke him* or *shake me*[64]), then easily obtain $7 \cdot 30103 = 210721 \rightarrow 10721$ (keeping only the five-digit mantissa) and add to conclude the calculation.

Usually this approximation is faster and more accurate than the one in the previous section. Practise both until you develop a preference.

Summary

Our general approach to calculating a common logarithm is to add and subtract known logarithms, typically of primes, to obtain the logarithm of the number itself if possible, otherwise the logarithm of a number very close to it. In the latter case, we make an adjustment using an estimate of $\log(1+x)$.

---

63 For the same reason, $\log 5 = 1 - \log 2 = 0.69897$ is full of large digits that tend to be inconvenient for calculation. Instead of adding $\log 5^n$ directly in the course of a calculation, I usually subtract $\log 2^n$; since the characteristic will be set right at the end, this amounts to adding $n - \log 2^n = \log 10^n - \log 2^n = \log(\frac{10}{2})^n = \log 5^n$.

64 My apologies for the violence; I struggled to think of better phrases.

Chapter 19

Exponentials

PERHAPS YOU WISH TO compute 10^x and e^x for their own sake. Often, however, they are needed as antilogarithms at the end of calculations involving logarithms (for which see the next chapter). Except in the easy cases of 10 to an integral power and $e^0 = 1$, these functions have irrational values and are difficult to compute.[65] Through reversal of the procedure for logarithms and judicious use of estimation, however, they can be computed not only manually but also mentally.

The natural exponential function

Calculation of e^x, the natural exponential function, is facilitated by a knowledge of a few convenient values given below.

Each multiple of $e^{2.3026}$ is a multiple of 10. Of the other values in the table, $e^{0.7}$ and $e^{1.1}$ are the most important and the best to remember, along with $e^1 = e$ itself.

---

65 In 1882, the German mathematician Ferdinand von Lindemann proved that e^x is transcendental if x is an algebraic (*i.e.*, non-transcendental) number other than 0. One implication of this is that π is transcendental. This follows from the fact, proven in the appendix on Taylor series, that $e^{i\pi} = -1$. Since -1 is algebraic, $i\pi$ is transcendental; but i is also algebraic, so π is transcendental. So is e, because 1 is algebraic and consequently $e^1 = e$ is transcendental. For this reason, the digits of π and e go on forever, without forming a repeating pattern. In addition, neither π nor e is the solution of any algebraic equation.

$e^{2.3026}$	10
$e^{-2.2}$	0.11080
$e^{0.7}$	2.01375
e^{1}	2.71828
$e^{1.1}$	3.00417
$e^{1.8}$	6.04965

For values of x close to 0, the equation

$$e^x = 1 + x + \frac{x^2}{2} + \frac{x^3}{2 \cdot 3} + \frac{x^4}{2 \cdot 3 \cdot 4} + \frac{x^5}{2 \cdot 3 \cdot 4 \cdot 5} + \cdots,$$

which gives the *Taylor series*[66] for e^x, affords a practical method of calculation. For example,

$$\begin{aligned} e^{0.1} &= 1 + 0.1 + \frac{0.1^2}{2} + \frac{0.1^3}{6} + \cdots \\ &\approx 1 + 0.1 + \frac{0.01}{2} + \frac{0.001}{6} \\ &\approx 1 + 0.1 + 0.005 + 0.00017 \\ &= 1.10517. \end{aligned}$$

Similarly,

$$\begin{aligned} e^{-0.1} &= 1 - 0.1 + \frac{0.1^2}{2} - \frac{0.1^3}{6} + \cdots \\ &\approx 1 - 0.1 + 0.005 - 0.00017 \\ &= 0.90483. \end{aligned}$$

When x is sufficiently small (whether positive or negative), the terms beyond the first two are so small that they can be ignored, and the Taylor series collapses to this simple form:

$$e^x \approx 1 + x.$$

66 Taylor series are explained in an appendix.

We can combine these various techniques, and the fact that $e^{x+y} = e^x e^y$, to calculate the exponentials of numbers up to ±6 or so, either roughly or accurately. For example,

$$\begin{aligned} e^{3.3} &= e^{2.3+1} \\ &= e^{2.3} e^{1} \\ &\approx 10e \\ &\approx 27.2. \end{aligned}$$

This rough but very quick estimate exploits the fact that $e^{2.3026} \approx 10$. For greater accuracy, we could proceed as follows, taking advantage of the fact that $e^{-0.0026} \approx 1 - 0.0026$. All six digits of the answer are correct, but we are entitled to only five significant digits, so we round the answer to 27.113.

$$\begin{aligned} e^{3.3} &= e^{2.3+1} \\ &= e^{2.3026+1-0.0026} \\ &= e^{2.3026} e^{1} e^{-0.0026} \\ &\approx 10e(1 - 0.0026) \\ &\approx 27.1828(1 - 0.0026) \\ &\approx 27.1828 - 27 \cdot 0.0026 \\ &= 27.1828 - 0.0702 \\ &\approx 27.1126. \end{aligned}$$

Note that it is easier to leave $e^{-0.0026}$ in the form $1 - 0.0026$ for multiplication than to express it as 0.0074.

We could calculate e^3 in this manner:

$$\begin{aligned} e^{3} &= e^{1.8+1.1+0.1} \\ &= e^{1.8} e^{1.1} e^{0.1} \\ &\approx 3.00417 \cdot 6.04965 \cdot 1.10517 \\ &\approx 6.05(3 + 0.14\%)(1 + 10.5\%) \end{aligned}$$

$$
\begin{aligned}
&= (18.15 + 0.14\%)(1 + 10.5\%) \\
&\approx (18.15 + 0.025)(1 + 10.5\%) \\
&\approx 18.175 + 10.5\% \\
&\approx 18.175 + 1.818 + 0.090 \\
&\approx 20.083 \\
&= 20.08 \text{ to four digits.}
\end{aligned}
$$

The correct answer is 20.086. Shortcuts in computation are responsible for the error. This superior method of calculating e^3 avoids the inaccuracy:

$$
\begin{aligned}
e^3 &= e^{2.3+0.7} \\
&= e^{2.3026+0.7-0.0026} \\
&= 10 \cdot 2.01375(1 - 0.0026) \\
&= 20.1375 - 0.26\% \\
&\approx 20.1375 - 0.0524 \\
&= 20.0851.
\end{aligned}
$$

The correct answer, to six digits, is 20.0855.

For a more difficult exponential, such as e^5, we must choose between accuracy and ease. A coarse estimate can be had quickly:

$$
\begin{aligned}
e^5 &= e^{0.7\cdot 7+0.1} \\
&= (e^{0.7})^7 e^{0.1} \\
&\approx 2.01375^7(1 + 0.10517) \\
&\approx (2 \cdot 1.007)^7(1 + 0.105) \\
&\approx 2^7(1 + 0.05)(1 + 0.105) \\
&\approx 128 + 7 + 13 \\
&\approx 148.
\end{aligned}
$$

Here we use the approximate binomial expansion

$$(1+x)^n \approx 1+nx,$$

good for small values of x, to obtain $1.007^7 \approx 1 + 0.007 \cdot 7 = 1.049 \approx 1.05$.[67]

Easier still:

$$\begin{aligned} e^5 &= e^{2.3\cdot 2+0.4} \\ &= (e^{2.3})^2 e^{0.4} \\ &\approx 10^2 \left(1 + 0.4 + \frac{0.4^2}{2}\right) \\ &= 100(1 + 0.4 + 0.08) \\ &= 100(1.48) \\ &= 148. \end{aligned}$$

This method calculates the Taylor series for e^x to three terms. Since already $e^{4.6} \approx 100$ is inaccurate, even the third digit of the answer is questionable—but it happens to be correct.

To calculate it more accurately, I would switch to $10^{5 \log e} \approx 10^{5\cdot 0.43429} = 10^{2.17145}$. If forced to use e^x directly, I would sweat it out in this clumsy (if arguably clever) manner, knowing that I would not gain much for my pains:

$$\begin{aligned} e^5 &= e^{1.1\cdot 3+0.7+1} \\ &= (e^{1.1})^3 e^{0.7} e \\ &\approx 3.00417^3 \cdot 2.01375 \cdot 2.71828 \\ &\approx (3+0.004)^3 \cdot 2.71828 \cdot 2.01375 \\ &\approx (3^3 + 3\cdot 3^2 \cdot 0.004) \cdot 2.71828 \cdot 2.01375 \\ &= 27.108 \cdot 2.71828 \cdot 2.01375 \\ &\approx (2.71^2 \cdot 10 + 0.3\%)2.01375 \end{aligned}$$

67 See the appendix on the binomial theorem.

$$\begin{aligned} &\approx (73.441 + 0.220)2.01375 \\ &\approx 73.661 \cdot 2 + 0.7\% \\ &\approx 147.32 + 1.03 \\ &= 148.35. \end{aligned}$$

All that toil scarcely generated a fourth correct digit: $e^5 = 148.41$. I include this tedious calculation not to recommend this decidedly inadvisable method of computing e^5 but to illustrate how a capable and thoughtful mental arithmetician can achieve reasonable accuracy even when faced with a forbidding tangle of products and powers of numbers many digits long.[68] If I do say so myself, exploiting the coïncidence that $3.004^3 \approx 10e$ was an elegant *coup de grâce.*

I seldom calculate e^x. If you expect to calculate it frequently, you should memorise the natural logarithms of a few easily multiplied numbers, especially primes, as follows:

$e^{0.69315}$	2
$e^{1.09861}$	3
$e^{1.60944}$	5
$e^{1.94591}$	7
$e^{2.39790}$	11

Because $\ln 2 = 0.69315$, we know that $e^{0.69315} = 2$, and so with the other values in the table. Armed with this knowledge, we can then calculate e^5 in the following manner:

$$e^5 = e^{2 \cdot 2.39790 + 0.20420}$$

68 The substitution $(3 + 0.004)^3 \approx 3^3 + 3 \cdot 3^2 \cdot 0.004$ follows from the binomial theorem, which is explained in an appendix.

$$
\begin{aligned}
&= \left(e^{2.39790}\right)^2 e^{0.20420} \\
&= 11^2 e^{0.20420} \\
&\approx 121\left(1 + 0.20420 + \frac{0.20420^2}{2} + \frac{0.20420^3}{6}\right) \\
&\approx 121\left(1.20420 + \frac{0.04162 + 0.00008}{2} + \frac{0.0086}{6}\right) \\
&\approx 121(1.20420 + 0.02085 + 0.00143) \\
&\approx 121(1.22648) \\
&\approx 11 \cdot 11(1.2265) \\
&\approx 11 \cdot 13.4915 \\
&\approx 148.406 \\
&= 148.41 \text{ to five digits.}
\end{aligned}
$$

In fact, $e^5 \approx 148.413$, so all five digits of the answer are correct.

The powers of 0.20420 need be calculated to only five decimal places—not even that many, because division by 2 or 6 tends to cause the last digit to fall away. Indeed, even $e^{0.20420}$ can be calculated to only four digits because it is multiplied by 121, which causes the last digits to lose significance. That is why I reduced the calculation to four decimal places towards the end. To five decimal places, $e^{0.20420} = 1.22654$, so my calculation was off by 0.00006, a greater error than I like; still, it was accurate to five digits, as was the final answer.

To calculate 0.20420^2, I obtained $204^2 = 41\,616$ and adjusted the decimal point; then I added $0.0008 = 2 \cdot 0.2 \cdot 0.0002$ as a correction because $(0.204 + 0.0002)^2 = 0.204^2 + 2 \cdot 0.204 \cdot 0.0002 + 0.0002^2$, of which only the second term need be retained (and only approximately). For 0.2042^3, my estimate was deliberately coarse: $0.2042^3 = 0.2042^2 \cdot 0.2042 \approx 0.042(0.2 + 0.004) = 0.0084 + 0.0002 = 0.0086$. No greater accuracy was needed, for the reasons stated above, so I did not waste effort.

Even simpler than the calculation above, but harder to notice: $e^5 = e^{2\cdot 1.94591+1.09861+0.00957} = 7^2 \cdot 3 \cdot e^{0.00957} = 147e^{0.00957}$. The approximation $e^{0.00957} \approx 1.00957 = 1.01 - 0.00043$, when multiplied by 147, gives $e^5 \approx 148.41$. The actual value of $e^{0.00957}$ is 1.00962 to five decimal places; it makes no difference in this calculation, where multiplication by 147 causes the last digit of the mantissa to lose significance.

These examples serve to illustrate how easily and accurately one can calculate e^x with just a few memorised values.

Powers of 10

Often the need to compute 10^x as an antilogarithm arises in calculations that use common logarithms. As with e^x, we find a known logarithm that is as close as possible to x and adjust for the difference. Significantly, the characteristic can be detached and incorporated at the end.

Suppose that we want 10^π. That is approximately $10^{3.14159}$. The characteristic, 3, can be ignored for the moment; we consider only the mantissa, 14159,[69] and try to find a known common logarithm close to it (whether above or below), preferably leaving a difference within ± 0.01.

We may know that $\log 13 = 11394$. That, however, is not close enough to 14159. Is $\log 15$ closer? It is $\log 3 + \log 5 \rightarrow \log 3 - \log 2 = 47712 - 30103$. The first two digits of the logarithm are 17, so already we see that it is too high. Try $\log 14$ in-

69 These exercises follow the usual practice of maintaining only the mantissa and discarding anything that affects only the characteristic. The notation is sloppy and incorrect, but it does reflect the thinking of a practical and busy mental calculator, who has no time to worry about formalities that do not affect the substance of the work. For instance, $\log 15$ in this exercise is taken to be $\log 3 - \log 2$. That is incorrect: $\log 15 = \log 30 - \log 2$. The factor of 10, however, affects only the characteristic and should be disregarded for efficient calculation of the mantissa.

stead: $\log 14 = \log 2 + \log 7 = 30103 + 84510 = 14613$. It is too high by 00454 (14613 − 14159), a difference small enough (less than 01000) that we can apply the estimate

$$e^x \approx 1 + x$$

in the form

$$10^x \approx 1 + 2.3023x,$$

which is based on the fact that $\ln 10 = \frac{1}{\log e} \approx 2.3023$. (Recall that the correct figure, to five digits, is 2.3026. In calculations of logarithms, the inaccuracy in the last digit tends to fall into insignificance because of multiplication by a number that is often less than 0.1. In calculations of exponentials, however, the substitution of 2.3023 sometimes does create inaccuracy in the last digit. We should therefore correct for the error, probably by adding a small amount to the last digit after multiplying by 2.3023.)

We calculate

$$\begin{aligned} 10^{0.14159} &= 10^{0.14613-0.00454} \\ &= 10^{0.14613}10^{-0.00454} \\ &\approx 14(1 - 2.3023 \cdot 0.00454). \end{aligned}$$

The multiplication is less difficult than it appears, in part because $2.3023 = 23 \cdot 0.1001$, a fact that enables us to multiply by 23 and then add the result shifted three decimal places to the right. Because we expect no more than five digits' accuracy, we need not compute the product exactly. Multiplying each digit in 454 by 23, we get $9200+1150+92 = 10442$ (or, in a more streamlined notation, $[92, 115, 92] = [103, 14, 2] = [104, 4, 2] = 10442$). Since 23 is too high by a factor of 10, we take instead $\frac{1}{10}$ of 10442, or 01044. That is $2.3 \cdot 0.00454$. To obtain $2.3023 \cdot 0.00454$, we shift 01044 three digits to the right and add it. Shifting it three digits to the right leaves only the 1, so we have $2.3023 \cdot 0.00454 \approx 01044 + 1 = 01045$. Therefore,

$$\begin{aligned}10^{0.14159} &\approx 14(1 - 2.3023 \cdot 0.00454)\\ &\approx 14 - 14 \cdot 0.01045\\ &= 14 - 0.14630\\ &= 13.85370\\ &= 13.854 \text{ to five digits.}\end{aligned}$$

Finally, we must account for the characteristic, 3. The answer is on the order of $10^3 = 1\,000$; therefore, it is 1 385.4. All five digits are correct.[70]

Finding an approximate answer in the form of a known logarithm is the greatest challenge. Knowledge of the logarithms of prime numbers is especially valuable, since they can be used to generate the logarithms of composite numbers.

❧

To raise 10 to a negative power, add an integer sufficient to make the power positive; this approach affects only the characteristic. For $10^{-2.71828}$, we start by calculating $10^{3-2.71828} = 10^{0.28172}$. Knowing that $\log 19 = 27875$, we can proceed immediately. If we do not know that logarithm, however, we may see that neither $\log 20 = 30103$ nor $\log 18 = 2\log 3 + \log 2 = 95424 + 30103 = 25527$ is close enough; we may then compute either $\log 18.9 = \log(9 \cdot 21) = 3\log 3 + \log 7 = 27646$ or, to simplify subsequent multiplication, $\log 19$ directly:

$$\begin{aligned}\log 19 &= \log(20 - 1)\\ &= \log 20\left(1 - \frac{1}{20}\right)\end{aligned}$$

70 That is, $10^{3.14159} = 1\,385.4$ to five digits. The value of 10^{π} is about 1 385.5.

$$
\begin{aligned}
&\approx \log 20 + 0.4343 \cdot \frac{-1/20}{1 - 1/(2 \cdot 20)} \\
&\approx \log 20 + 0.4343 \cdot \frac{-2}{39} \\
&\approx \log 20 - 2 \cdot 0.011135 \\
&= \log 20 - 0.02227 \\
&= 30103 - 02227 \\
&= 27876.
\end{aligned}
$$

This estimate of log 19 is slightly higher than the actual value, 27875, given above. Continuing (well aware that the decimal point is in the wrong place for most of this calculation):

$$
\begin{aligned}
10^{0.28172} &= 10^{0.27875+0.00297} \\
&= 10^{0.27875} 10^{0.00297} \\
&\approx 19\big(1 + 2.3023(0.00297)\big) \\
&= 19\big(1 + 2.3023(0.003 - 0.00003)\big) \\
&\approx 19\big(1 + (0.00691 - 0.00007)\big) \\
&\approx 19 \cdot 1.00684 \\
&\approx 19.1300 \\
&\rightarrow 1.9130 \text{ to five digits.}
\end{aligned}
$$

That gives us $10^{0.28172}$. Since $10^{-2.71828} = 10^{0.28172-3}$, we multiply by 10^{-3} and conclude that $10^{-2.71828} \approx 0.001\,913\,0$. Indeed, all digits of our answer are correct.

Powers of other bases

In general, exponentials of bases other than e and 10 are most easily calculated using common logarithms; the technique is explained in the following chapter. If you frequently compute a function such as 2^x, however, you might design a method that incorporates memorised logarithms to the relevant base, analogously to the ones presented above.

Hyperbolic functions and their inverses

The trigonometric, or circular, functions are based on the unit circle; their analogues the hyperbolic functions are based on the unit hyperbola.[71]

The *hyperbolic sine* (sinh), *hyperbolic cosine* (cosh), and *hyperbolic tangent* (tanh) are defined as follows:

$$\sinh x = \frac{e^x - e^{-x}}{2};$$
$$\cosh x = \frac{e^x + e^{-x}}{2};$$
$$\tanh x = \frac{e^x - e^{-x}}{e^x + e^{-x}} = \frac{e^{2x} - 1}{e^{2x} + 1}.$$

Their inverses are defined using the natural logarithm:

$$\operatorname{arsinh} x = \ln\left(x + \sqrt{x^2 + 1}\right);$$
$$\operatorname{arcosh} x = \ln\left(x + \sqrt{x^2 - 1}\right);$$
$$\operatorname{artanh} x = \frac{1}{2}\ln\left(\frac{1 + x}{1 - x}\right).$$

Rarely do I have the occasion to calculate hyperbolic or inverse hyperbolic functions. When I do, I simply apply the definitions.

71 The relationship between the two sets of functions is mediated through the complex numbers. From the fact that $e^{ix} = \cos x + i \sin x$, we have $\sin(ix) = i \sinh x$; $\cos(ix) = \cosh x$; $\tan(ix) = i \tanh x$; $\sinh(ix) = i \sin x$; $\cosh(ix) = \cos x$; and $\tanh(ix) = i \tan x$. Could it be said that the hyperbola is the circle of the complex numbers?

The second form for tanh x is the more convenient, since it requires the calculation of one exponential only (we certainly prefer not to calculate two of them!). Anyone who needs the hyperbolic tangent, however, will find a good mental algorithm in an impressive paper that is freely available on the Internet.[72]

---

72 Ron Doerfler, «Fast Approximation of the Tangent, Hyperbolic Tangent, Exponential and Logarithmic Functions» at 8–12. See the bibliography.

Chapter 20

Computational applications of logarithms

LOGARITHMS were invented not for their own sake but as a computational aid, especially to simplify calculations involving multiplication or division of trigonometric functions. Beyond being presented in books of tables, logarithms found concrete implementation in the form of slide rules, which remained important as calculating instruments until the mid-1970s, when affordable electronic calculators displaced them. Most of the calculations formerly done by hand with logarithms are now relegated to calculators or computers, but some are still amenable to mental arithmetic, and the results can be impressive indeed, even if they are accurate to only two or three digits.

Estimating large numbers

How many digits are there in 3^{100}? Using the common logarithm, we can easily find out, and even obtain the first few digits of the answer.

Since $\log x^y = y \log x$, we have

$$\log 3^{100} = 100 \log 3 \approx 100 \cdot 0.47712 = 47.712.$$

Therefore,

$$3^{100} \approx 10^{47.712} = 10^{47}10^{0.712}.$$

We know that $10^{0.712}$ is between 1 and 10 (10^0 and 10^1). Since 10^{47} is a 1 followed by 47 zeroes, it is a 48-digit number, and so is its product with $10^{0.712}$. Thus 3^{100} has 48 digits.

Furthermore, the first few digits can be obtained from the mantissa of the logarithm. Since $\log 5 = 0.69897$ is just slightly less than 0.712, we can say that 3^{100} is a bit more than 5×10^{47}. Two memorised logarithms and a very simple calculation were all that we needed to obtain that estimate.

For a better estimate still, we might try $\log 5.1 \rightarrow \log 17 + \log 3 = 23045 + 47712 \rightarrow 0.70757$, which is slightly less than 0.712. Thus we can say of 3^{100} «It's a bit more than 5.1×10^{47}»—an impressive result obtained without much effort.[73] (Calculation of $10^{0.712}$ would yield 5.15, to three significant digits; the answer would therefore be 5.15×10^{47}.)

Difficult powers

How would you calculate $2^{3.42}$? Because it is equal to $2^{342/100}$, you could calculate it as $\sqrt[100]{2^{342}}$ or as $\left(\sqrt[100]{2}\right)^{342}$. I certainly am not volunteering for the task.

Logarithms to the rescue! They usually offer the most practical way to calculate difficult powers such as this one. Indeed, as is mentioned in the first chapter, irrational powers are *defined* in terms of the natural logarithm and the exponential function: specifically, $x^y = e^{y \ln x}$.

$$\begin{aligned} 2^{3.42} &= 10^{\log 2^{3.42}} \\ &= 10^{3.42 \log 2} \end{aligned}$$

73 If you have a taste for showing off, you may say instead «It's a bit more than 510 quattuordecillion».

$$\approx 10^{3.42 \cdot 0.30103}$$
$$\approx 10^{1.02952}.$$

The logarithm, 1.02952, is slightly greater than 1; therefore, the power is slightly greater than $10^1 = 10$. To estimate it roughly, we can rely on the fact that 0.02952 is small; its antilogarithm is therefore approximately $1 + 2.3023 \cdot 0.02952 \approx 1.07$. Multiplying by 10 (because of the characteristic 1) yields 10.7 as the approximate answer.

For greater accuracy, we could calculate $10^{1.02952}$ by taking log 1.07, already found, as a good estimate:

$$\begin{aligned}
\log 1.07 \to \log 107 &= \log 100 \left(1 + \frac{7}{100}\right) \\
&\approx \log 100 + 0.4343 \cdot \frac{7/100}{1 + 7/(2 \cdot 100)} \\
&= 00000 + 0.4343 \cdot \frac{14}{207} \\
&\approx 14 \cdot 0.002098 \\
&= 0.02937.
\end{aligned}$$

(The actual value is 0.02938.) This logarithm differs from the exponent by 00015. The answer is then

$$\begin{aligned}
10.7(1 + 2.3023 \cdot 0.00015) &\approx 10.7 + 10.7 \cdot 0.00035 \\
&\approx 10.7 + 0.0037 \\
&= 10.704.
\end{aligned}$$

The correct value, to five digits, is 10.703. The slight error in our estimate of log 1.07 is responsible for the inaccuracy in the last digit.

Difficult roots

As was seen in the chapter on roots, direct calculation of roots other than the square root is usually cumbersome, especially if three or more accurate digits are needed. Like difficult powers, difficult roots can be calculated with logarithms. After all, a root is just a power: $\sqrt[n]{x} = x^{1/n}$.

Direct calculation of $\sqrt[17]{234\,000\,000}$ would be troublesome indeed. With logarithms, the extraction of the root still takes work, but it can be done mentally:

$$\begin{aligned}\sqrt[17]{234\,000\,000} &= 10^{\log \sqrt[17]{234\,000\,000}} \\ &= 10^{(\log 234\,000\,000)/17}.\end{aligned}$$

The logarithm can be calculated as follows:

$$\begin{aligned}\log 234\,000\,000 &= \log\left(234 \cdot 10^6\right) \\ &= \log\left(2 \cdot 3^2 \cdot 13 \cdot 10^6\right) \\ &= \log 2 + 2\log 3 + \log 13 + \log 10^6 \\ &\approx 0.30103 + .95424 + 1.11394 + 6 \\ &= 8.36921.\end{aligned}$$

We can now finish the problem:

$$\begin{aligned}10^{(\log 234\,000\,000)/17} &\approx 10^{8.36921/17} \\ &\approx 10^{0.49231}.\end{aligned}$$

Because 0.49231 is a bit greater than $\log 3 = 0.47712$, we know that $\sqrt[17]{234\,000\,000}$ is a bit greater than 3. For a better approximation, consider log 3.2:

$$\log 3.2 = \log \frac{32}{10}$$

$$
\begin{aligned}
&= \log 2^5 - \log 10 \\
&= 5 \cdot 0.30103 - 1 \\
&= 0.50515.
\end{aligned}
$$

Thus 0.49231 is between log 3 and log 3.2—a bit closer to the latter. Linear interpolation makes a quick estimate possible. The distance between 3 and 3.2 is 0.2; to find how much greater $10^{0.49231}$ is than 3, we scale 0.2 as follows:

$$
\begin{aligned}
0.2 \cdot \frac{0.49231 - 0.47712}{0.50515 - 0.47712} &\approx 0.2 \cdot \frac{0.015}{0.028} \\
&= 0.2 \cdot \frac{15}{28} \\
&= 0.1 \cdot \frac{15}{14} \\
&= 0.1 \cdot 1.07 \\
&= 0.107.
\end{aligned}
$$

Hence $\sqrt[17]{234\,000\,000}$ is approximately 3.107.

Using instead the more accurate methods for antilogarithms, we obtain $10^{0.49231} = 3.106\,8$. First, $\log 31 = \log 30(1 + \frac{1}{30}) \approx \log 30 + 0.4343 \cdot \frac{2}{61} = 47712 + 01424 = 49136$. This estimate falls short of the exponent by only 00095. Next, $31(1 + 2.3023 \cdot 0.00095) \approx 31(1 + 0.00219) \approx 31.068$. Adjusting the decimal point to reflect the characteristic yields 3.106 8, the correct value of $\sqrt[17]{234\,000\,000}$ to five digits.

Chapter 21

Financial calculations

WEALTH GREAT OR SMALL requires arithmetic for its management. You will impress your accountant or your banker by performing in your head the calculations in this section. Most people in the financial realm, if they can perform these calculations at all, rely on special financial calculators; few would even know how to perform them on their own, and fewer still could do so mentally.

Compound interest

If you draw your interest out as soon as the bank pays it, you receive *simple interest*. The amount is easy to calculate: we just multiply the interest rate by the amount invested. For example, \$1 000 invested at 5% per year will yield \$50 in annual interest.

Great things happen, however, if instead of drawing your interest out you add it to your investment: you will then receive interest on top of the interest already paid. This is known as *compound interest*, and it grows much more rapidly than simple interest. The growth is so rapid that young people who regularly save and invest their money are likely to be well rewarded when they retire.

If P is the principal (the initial amount invested) and compound interest is paid at the rate of r per period (expressed as a decimal fraction, so that $r = 0.05$ for 5% interest), at the end of the first period the investor will have P plus interest of rP, for a total of $P + rP = P(1 + r)$. Thus the payment of interest multiplies the principal by $1 + r$.

In the second period, the principal will include the interest previously paid: it will be $P(1 + r)$. As was just seen, the next payment of interest will multiply the principal by $1 + r$, yielding $P(1+r)(1+r) = P(1+r)^2$. Similarly, the third period will bring the total to $P(1 + r)^3$, and in general at the end of period t the total will be

$$P(1 + r)^t.$$

This is a form of *exponential growth*, because the factor $1 + r$ is raised to the power of the number of periods. Exponential growth sooner or later becomes very rapid. After 40 years at 5% per annum, compound interest will have multiplied principal by 1.05^{40}, which is about 7.04, thereby turning a \$1 000 investment into \$7 000. Simple interest would have multiplied the money by only $1 + 0.05 \cdot 40 = 3$.

The great seventeenth-century Swiss mathematician Jacob Bernoulli wondered what would happen if compound interest were paid continuously, not just at the end of the period. With n equal intervals in the period, the rate of interest paid per interval would be r/n, and the amount of money after n periods would be

$$P\left(1 + \frac{r}{n}\right)^n.$$

For continuous compounding, he sought to know whether that value approaches a limit as n increases towards infinity—and the answer is yes. In the calculus, this limit is expressed as

$$\lim_{n\to\infty} P\left(1 + \frac{r}{n}\right)^n.$$

If the interest rate is 100%, then $r = 1$. Taking P to be 1 for simplicity's sake, we obtain

$$\lim_{n\to\infty}\left(1+\frac{1}{n}\right)^{n}.$$

Bernouilli proved that this limit does exist, and that it lies between 2.5 and 3. The limit turns out to be the number e,[74] and Bernouilli's work on compound interest contains the first reference to it. Natural logarithms and their base turn up in surprising places.

Usually $(1+r)^t$ is most easily calculated with logarithms and exponentials. For values of r less than 0.1 (10%) or so, however, the *rule of 72* offers a quick and easy estimate of the time needed to double one's money. We want to know the value of t at which

$$(1+r)^t = 2.$$

Taking the natural logarithm of each side of this equation, with the fact that $\ln x^y = y\ln x$, yields

$$t\ln(1+r) = \ln 2.$$

Therefore,

$$t = \frac{\ln 2}{\ln(1+r)}.$$

Although the logarithm to any base would have worked just as well, the choice of the natural logarithm is especially helpful

74 Proving this requires a knowledge of the calculus. Let $f(x) = \ln x$; then $f'(x) = 1/x$, so $f'(1) = 1$. By the definition of the derivative, we also have $f'(1) = \lim_{\Delta x\to 0}\big(f(1+\Delta x) - f(1)\big)/\Delta x = \lim_{\Delta x\to 0}\frac{1}{\Delta x}\ln(1+\Delta x) = \lim_{\Delta x\to 0}\ln(1+\Delta x)^{1/\Delta x}$; therefore, $\lim_{\Delta x\to 0}\ln(1+\Delta x)^{1/\Delta x} = 1$. Because e^x is continuous, we have $\lim_{\Delta x\to 0}\exp\big(\ln(1+\Delta x)^{1/\Delta x}\big) = e^1 = e$. Therefore, $\lim_{\Delta x\to 0}(1+\Delta x)^{1/\Delta x} = e$, and equivalently $\lim_{n\to\infty}(1+1/n)^n = e$.

here because, as is shown in the chapter on logarithms, the following approximation holds for small values of r:

$$\ln(1+r) \approx r.$$

Thus

$$t \approx \frac{\ln 2}{r} \approx \frac{0.693}{r}.$$

If r is expressed as a percentage, $69.3/r$ gives the approximate number of periods until the original principal is doubled. We use $72/r$ instead, both because the substitution of r for $\ln(1+r)$ leads to a slight underestimate and particularly because 72 can be divided easily by many common interest rates.

Thus money invested at 8% interest will double in approximately $72/8 = 9$ years. At 4%, it will double in approximately $72/4 = 18$ years.

Amortisation

Loans such as mortgages are typically repaid through equal periodic payments. Each payment is applied to both principal and interest, but the allocation varies: since the balance of the loan declines as payments are made towards principal, the interest due also declines, and over time an increasing share of the monthly payment goes to principal. This process of repaying an interest-bearing loan through fixed periodic payments is known as *amortisation*.

For a loan whose initial principal is P, with interest of r (expressed as a decimal, not as a percentage) per period for n periods, the following equation gives the monthly payment A:

$$A = P \cdot \frac{r}{1-(1+r)^{-n}}.$$

The derivation of this equation appears in an appendix.

On a \$200 000 mortgage at 6% annual interest, with 30 years of monthly payments, the *monthly* interest, r, is $6\%/12 = 0.005$; the number of periods, n, is $30 \cdot 12 = 360$, because there are 12 monthly payments in a year. The amount of the monthly payment is

$$A = 200\,000 \cdot \frac{0.005}{1 - 1.005^{-360}} \approx 1\,199.10.$$

Financial calculators and Web sites that generate amortisation schedules confirm that the monthly payment is \$1 199.10.

Having learnt to calculate difficult powers using logarithms and exponentials, you could estimate the monthly payment in your head.[75] You would be forgiven, however, for hoping to learn of an easier method. We can now derive one that is better suited to mental calculation.

First, express the power in the denominator with the exponential function:

$$P \cdot \frac{r}{1 - (1 + r)^{-n}} = P \cdot \frac{r}{1 - e^{-n \ln(1+r)}}.$$

At realistic interest rates, r is small; therefore, $\ln(1 + r) \approx r$, and we have the approximation

$$P \cdot \frac{r}{1 - e^{-nr}}.$$

To obtain nr also in the numerator, we multiply the numerator by n and divide by n elsewhere:

$$\frac{P}{n} \cdot \frac{nr}{1 - e^{-nr}}.$$

75 While writing this section, I computed a coarse mental estimate: I obtained 0.84 for the denominator of the fraction, then doubled numerator and denominator to obtain 0.01/1.68. The denominator is approximately 10/6; thus the fraction is approximately 6/1 000, so the monthly payment is approximately $200 \cdot 6$, or 1 200. My sloppy calculation was surprisingly accurate.

In the fraction on the right, we substitute $y = nr$ to obtain

$$\frac{y}{1 - e^{-y}},$$

which, using the technique of Taylor series from the differential calculus,[76] is approximately

$$1 + \frac{y}{2} + \frac{y^2}{12} - \frac{y^4}{720}.$$

Unless the interest rate is high (above 7% on a 30-year loan), we can drop the last term (the one containing y^4) without losing more than 1% accuracy or so. Dropping that term and substituting $x = y/2$ yields

$$1 + x + \frac{x^2}{3}.$$

We can therefore approximate the monthly payment rather accurately with the following formula, in which $x = nr/2$ (since $x = y/2$ and $y = nr$):

$$A \approx \frac{P}{n}\left(1 + x + \frac{x^2}{3}\right).$$

Let us test this formula by applying it to the mortgage described above. We still have $P = 200\,000$ and $n = 360$. To calculate x, we need not use n and r; we can get the same result

76 If you are well versed in the calculus, you can verify this truncated Taylor series for yourself by computing the first few derivatives of the function being approximated. (If you are rusty on Taylor series, see the relevant appendix.) Otherwise, please be satisfied with the explanation that a Taylor series uses the calculus to turn a difficult function into the sum of powers of its variable—y^0, y^1, y^2, y^3, and so on forever—, each of those powers being multiplied by a constant. Often the Taylor series is so good that it can be cut short without much loss of accuracy. Such is the case here, unless y is uncommonly large —probably because of an unusually high interest rate.

by using the original annual figures of 30 years at 6% = 0.06 interest. Thus $x = 30 \cdot 0.06/2 = 0.9$. The approximate monthly payment is

$$\frac{200\,000}{360}(1 + 0.9 + 0.27) \approx 1\,206,$$

or \$1 206 per month, compared with the exact value of \$1 199.10. The error is less than 0.6%, but the arithmetic is far easier than that of the exact formula.

Chapter 22

Trigonometric functions

EVALUATING the trigonometric functions is so tedious, for all but a few important angles, that people for millennia have relied on tabulated values. Calculators nowadays are even more convenient than tables, and noöne can be criticised for using them. The techniques in this chapter, however, yield very quick estimates of the principal trigonometric functions, albeit sometimes only to two or three digits' accuracy.

The trigonometric functions are difficult to calculate. First, their ranges are inconvenient. Both sine and cosine always fall between −1 and +1, inclusive, so errors are likely to be proportionally large. Tangent, by contrast, ranges over all real numbers for angles between $-90°$ and $90°$; near the extremes, it changes very rapidly. The graph below shows all three functions.

In addition, series for them tend to be inconvenient for mental calculation. It is true that

$$\sin x = x - \frac{x^3}{3!} + \frac{x^5}{5!} - \frac{x^7}{7!} + \cdots$$

and

$$\cos x = 1 - \frac{x^2}{2!} + \frac{x^4}{4!} - \frac{x^6}{6!} + \cdots,$$

in which $a!$, called the *factorial* of a, is the product of all integers from 1 to a (*e.g.*, $4! = 4 \cdot 3 \cdot 2 \cdot 1 = 24$). According to the value of x, however, one might need four or five terms for reasonable

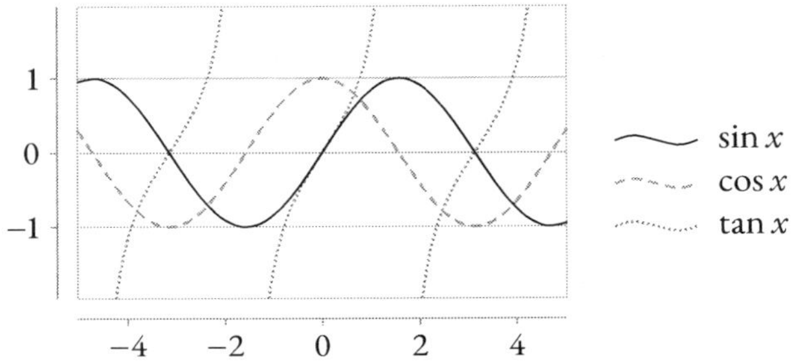

accuracy, and most of the terms are hard to calculate—especially because x is expressed in radians, not degrees.

Except where radians are specified, all trigonometric functions in this chapter use degrees. Where an error is given, it represents the correct value minus the approximation.

Conversion between radians and degrees

To convert degrees to radians, multiply by $\frac{\pi}{180}$, which is approximately 0.0175. To convert from radians to degrees, divide by the same number or, equivalently, multiply by $\frac{180}{\pi}$, which is approximately 57.3.

Neither these irrational numbers nor their decimal approximations are suitable for mental calculation. The fraction $\frac{401}{7}$ approximates $\frac{180}{\pi}$ to three digits and is convenient for both multiplication and division.[77]

Sine

The sine function repeats the same pattern every 360°; thus

$$\sin(x + 360n) = \sin x$$

77 See Doerfler, *Dead Reckoning* at 146, 172.

for every integer n. In addition, the segment from 0° to 90° is repeated in three inverted or mirror-image forms (90°–180°, 180°–270°, 270°–360°), because

$$\sin(x + 90) = -\sin(x - 90)$$

and

$$\sin(-x) = -\sin x.$$

Therefore, the values of $\sin x$ for every value of x between 0° and 90° can be used to generate all values of $\sin x$.

Since the graph of $\sin x$ from 0° to 180° resembles a parabola, we can try to fit a parabola to the sine wave over that interval. A parabola has the equation

$$y = ax^2 + bx + c$$

for some values of a, b, and c. We wish to find values for which the parabola approximates $y = \sin x$. To do so, we need to solve

$$ax^2 + bx + c = \sin x$$

for three values of x.

Since $\sin x = 0$ when $x = 0$, we obtain $c = 0$. Our equation therefore assumes this simpler form:

$$ax^2 + bx = \sin x.$$

Using the values 45 and 90 for x, we obtain the system of equations

$$2025a + 45b = \sin 45 = \frac{\sqrt{2}}{2}$$
$$8100a + 90b = \sin 90 = 1.$$

Solving these, we obtain the approximate values $a = -0.000102$ and $b = 0.0203$, which, for the sake of easier calculation, we simplify to $a = -\frac{1}{10\,000}$ and $b = \frac{1}{50}$. Therefore,

$$\sin x \approx \frac{x}{50} - \frac{x^2}{10\,000}$$
$$= 1 - \frac{(100 - x)^2}{10\,000}.$$

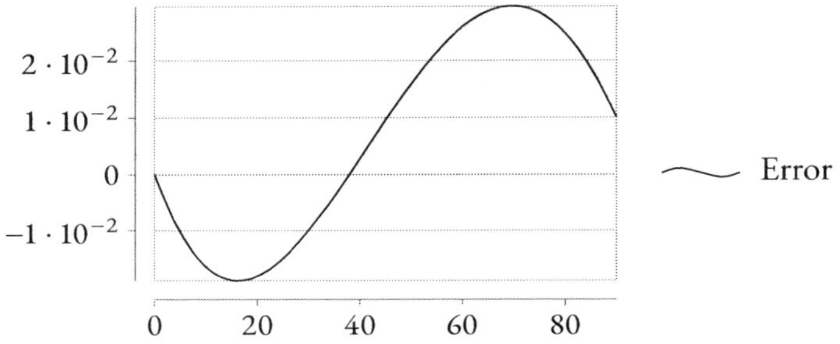

The error is 0 at 0°; it falls to −0.02 around 15°, rises to 0 again around 40°, rises to 0.03 around 70°, and falls to 0.01 at 90°. We can improve the accuracy of the calculation by making an appropriate adjustment for the error. For example, $\sin 10 \approx 1 - 0.81 = 0.19$; since the error is between 0 and −0.02, we might subtract about 0.015 to obtain 0.175. The correct value, to three digits, is 0.174. Similarly, $\sin 60 \approx 1 - 0.16 = 0.84$; the error is close to 0.03, so we might add 0.025 for an estimated value of 0.865, which is close to the correct value, 0.866.

Because we can expect only two digits' accuracy, we round these answers. Sometimes rounding throws the last digit off, for example, our estimate of sin 10 becomes 0.18, slightly different from the actual value of 0.17.

Knowledge of the squares of the two-digit numbers is helpful, but a coarse approximation suffices, since in any event the calculation will be good to only two decimal places.

Cosine

For the cosine, we have

$$\cos x = \sin(90 - x) \approx 1 - \frac{(x+10)^2}{10\,000}.$$

The error is the mirror image of that for $\sin x$, with the minimum of -0.02 around 75° and the maximum of 0.03 around 20°.

Tangent

The tangent function may appear untamable, ranging as it does over the entire set of real numbers for angles between −90° and +90°. Quotients of polynomials, however, can yield surprisingly good approximations.

The following approximation uses degrees. Although more involved than the ones for the sine and the cosine, it can be calculated mentally for values between 0° and 90°.

$$\tan x \approx \frac{x}{90 - x} \cdot \frac{990 - 4x}{630 + 4x}.$$

For x between 0° and 50° or so, the error is less than 0.005, so the approximation is good to two decimal places. Outside that

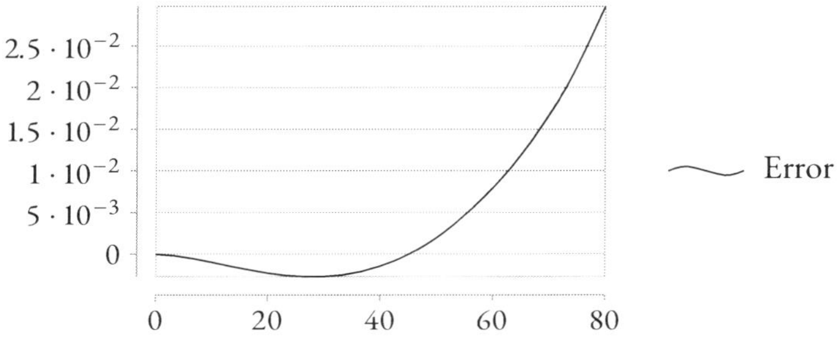

range, the approximation is still good. It can be improved, however, using the following identity:[78]

$$\tan x = \frac{1}{\tan(90 - x)}.$$

Thus for $x > 50$, we can instead calculate the reciprocal of the expression above with $90 - x$ substituted:

$$\tan(90 - x) \approx \frac{90 - x}{x} \cdot \frac{630 + 4x}{990 - 4x}.$$

The following example shows the calculation of tan 60. Note that 60° is outside the range in which this approximation is most accurate.

$$\begin{aligned}\tan 60 &\approx \frac{60}{30} \cdot \frac{750}{870} \\ &= 2 \cdot \frac{\not{3} \cdot \not{10} \cdot 25}{\not{3} \cdot \not{10} \cdot 29} \\ &= \frac{50}{29} \\ &\approx 1.72.\end{aligned}$$

The actual figure is 1.73.[79] We could correct for the error at 60° by adding 0.01 (see the graph) to obtain 1.73, the correct answer.

---

78 To understand why this identity is true, recall that the three angles of any triangle measure 180° in all. Therefore, since a right angle measures 90°, the two remaining angles of a right triangle together measure 90°, and each angle is 90° minus the other. The tangent of an angle is the opposite divided by the adjacent. Since the opposite side of one angle is the adjacent side of the other, and *vice versa*, the tangents of the two angles are reciprocals. Therefore, $\tan x$ is the reciprocal of $\tan(90 - x)$.

79 The exact value of tan 60 is $\sqrt{3}$.

To see how good the estimate is for $0 \leq x \leq 50$, try the following:

$$\begin{aligned}\tan 40 &\approx \frac{4\not{0}}{5\not{0}} \cdot \frac{83\not{0}}{79\not{0}} \\ &= 0.8\left(1 + \frac{4}{79}\right) \\ &\approx 0.8 \cdot 1.05 \\ &\approx 0.84.\end{aligned}$$

This result is correct: to three digits, $\tan 40 = 0.839$.

Better approximations

Unfortunately, better approximations come at a cost of greater computational complexity. The ones below work when the trigonometric function of x is known and δ is small relative to x:

$$\begin{aligned}\sin(x+\delta) &\approx \sin x + \left(\frac{\pi}{180}\cdot\delta\right)\cos x \\ \cos(x+\delta) &\approx \cos x - \left(\frac{\pi}{180}\cdot\delta\right)\sin x \\ \tan(x+\delta) &\approx \tan x + \left(\frac{\pi}{180}\cdot\delta\right)(\tan^2 x + 1).\end{aligned}$$

They do, however, require knowledge of the values of one or two trigonometric functions of an angle near the one in question. If δ is expressed in degrees, it must be converted to radians through multiplication by $\pi/180$.

The seventh-century Indian mathematician Bhāskara I discovered this approximation for values of x in degrees between 0 and 180:

$$\sin x \approx \frac{4x(180-x)}{40\,500 - x(180-x)}.$$

The absolute error is within ± 0.001; it reaches its minimum of -0.0016 around 10° and 170°, its maximum of 0.0014 around

50° and 130°. By making appropriate adjustments near the four peaks and valleys, one can expect almost three digits' accuracy.

This easily calculated approximation of the cosine is quite accurate for angles in degrees between 0 and 40: within ±0.0005 for most of those angles, the greatest error being −0.0011 in the vicinity of 30°.

$$\cos x \approx 1 - \frac{x(x+1)}{7\,000}.$$

To compute $\sin x$ for values of x between 40 and 90, apply the equation $\sin x = \cos(90 - x)$.

The following Padé approximant for $\tan x$ is highly accurate for values of x, in radians, from 0 to $\frac{\pi}{2}$:

$$\tan x \approx \frac{x - \frac{1}{9}x^3 + \frac{1}{945}x^5}{1 - \frac{4}{9}x^2 + \frac{1}{63}x^4} = \frac{945x - 105x^3 + x^5}{15(63 - 28x^2 + x^4)}.$$

For mental calculation, however, it is obviously impractical—not least because x must be given in radians.

Small angles

For small values of x, expressed in degrees, the following approximations are good within 1%:

$$\sin x \approx \left(\frac{\pi}{180}\right)x \qquad \text{for } |x| < 8°$$

$$\cos x \approx 1 \qquad \text{for } |x| < 14°$$

$$\tan x \approx \left(\frac{\pi}{180}\right)x \qquad \text{for } |x| < 10°$$

If x is expressed in radians, omit the factor $\pi/180$ to obtain the much simpler approximations $\sin x \approx x$ and $\tan x \approx x$.

Arcsine

The inverse of my approximation

$$\sin x \approx 1 - \frac{(100 - x)^2}{10\,000}$$

gives the following expression for the arcsine in degrees for angles between 0° and 90°:

$$\arcsin x \approx 100 - 100\sqrt{1 - x}.$$

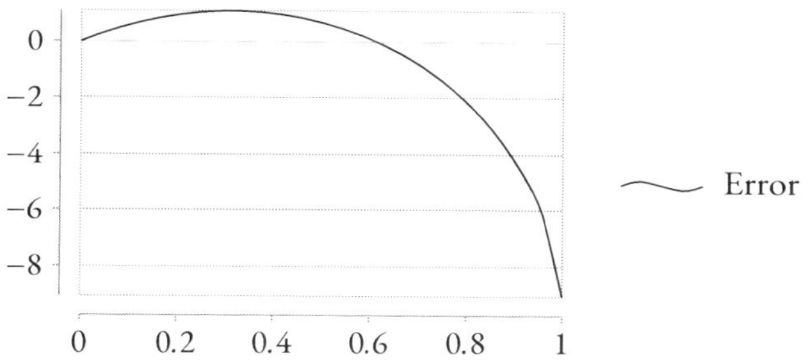

For values of x from 0 to 0.7 or so, the error is small. It is unfortunately large for $-1 \leq x < 0$; values in that range are best computed as $\arcsin |x|$. It is also large when $x > 0.8$.

For example, what is arcsin 0.6? We have $100 - 100\sqrt{0.40}$. Since $3\,969 = 63^2$, the square root is slightly greater than 0.63. Therefore, the estimate of the arcsine is slightly less than 37°. The graph suggests that the error at 0.6 is close to 0, and indeed the actual arcsine, to three digits, is 36.9°.

The following approximation of the arcsine in degrees is the inverse of the approximation of the sine given by Bhāskara I:

$$90 - \frac{180\sqrt{4 - 3x - x^2}}{4 + x}.$$

Arccosine

The inverse of my approximation

$$\cos x \approx 1 - \frac{(x+10)^2}{10\,000}$$

gives the following expression for the arccosine in degrees for angles between 0° and 90°:

$$\arccos x \approx 100\sqrt{1-x} - 10.$$

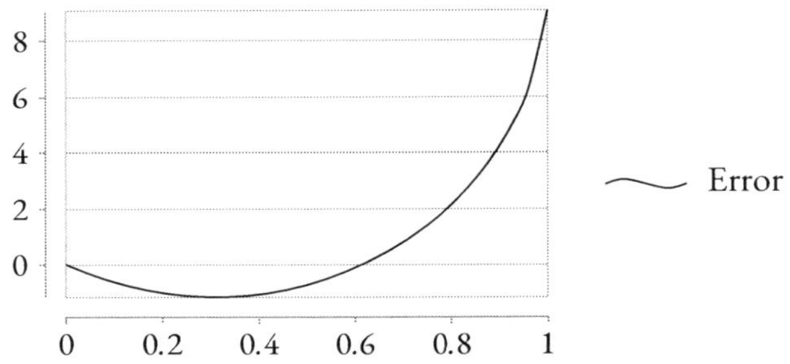

For values of x from 0 to 0.7 or so, the error is at most ±1°, and typically less.

A good approximation of the arcsine can yield the arccosine because

$$\arcsin x + \arccos x = 90;$$

therefore,

$$\arccos x = 90 - \arcsin x.$$

Arctangent

The following formula gives the approximate arctangent in degrees for angles between −45° and 90°:[80]

$$\arctan x \approx \begin{cases} \frac{401}{7} \cdot \frac{x}{1 + 0.28x^2} & \text{if } -1 \leq x \leq 1, \\ 90 - \frac{401}{7} \cdot \frac{x}{x^2 + 0.28} & \text{if } x > 1. \end{cases}$$

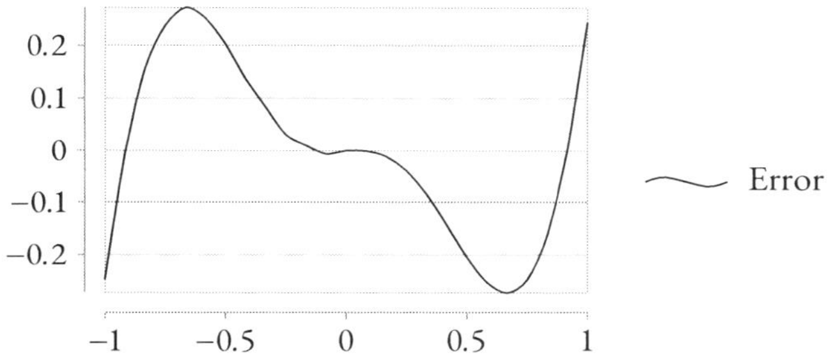

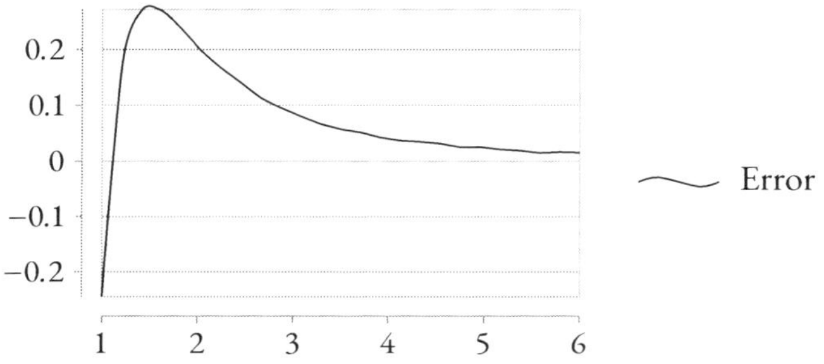

Throughout the domain of this approximation, therefore, the maximum error is around ±0.3, in the vicinity of ±0.7. At most values of x, the error is much smaller.

80 See Doerfler, *Dead Reckoning* at 161–63. The formula comes from a book by M Abromowitz and IA Stegun.

For example,

$$\begin{aligned}
\arctan 0.5 &\approx \frac{401}{7} \cdot \frac{0.5}{1 + 0.28 \cdot 0.5^2} \\
&= \frac{401}{7} \cdot \frac{1}{2 + 0.14} \\
&= \frac{401}{14.98} \\
&\approx \frac{401}{15} + 0.15\% \\
&= \frac{80.2}{3} + 0.15\% \\
&\approx 26.7 + 0.15\% \\
&\approx 26.74 \\
&\approx 26.7 \text{ to three digits.}
\end{aligned}$$

The graph above suggests subtracting approximately 0.2 to obtain 26.5°. The actual arctangent, to four digits, is 26.57°.

Another example:

$$\begin{aligned}
\arctan 3 &\approx 90 - \frac{401}{7} \cdot \frac{3}{3^2 + 0.28} \\
&\approx 90 - \frac{401}{7} \cdot \frac{1}{3.09} \\
&\approx 90 - \left(\frac{401}{21} - 3\%\right) \\
&\approx 90 - (19.1 - 3\%) \\
&\approx 90 - 18.53 \\
&\approx 71.47 \\
&\approx 71.5 \text{ to three digits.}
\end{aligned}$$

Adding 0.1 to offset the error, we obtain 71.6°, which is correct: to four digits, $\arctan 3 = 71.57°$.

Chapter 23

Showmanship

ZEAL FOR mental arithmetic is its own reward, but who can be blamed for enjoying the admiration of others? Short of making a circus of yourself, you may reasonably flaunt your hard-won skills; at least you may display them to best advantage. Here are some suggestions for giving 'em the old razzle-dazzle.[81]

Speed, accuracy, and perceived complexity or difficulty make a performance of mental arithmetic particularly impressive. For a general audience, squares of three-digit numbers may balance these characteristics effectively. A more mathematically sophisticated audience may like to see you calculate square roots to six figures. Calculations such as exponentials or squares of five-digit numbers, if performed correctly, will earn cachet for accuracy and complexity but not for speed.

You can relieve the burden on your memory by contriving to have some problems presented in written form. If you perform in front of a group, you may write them on a blackboard yourself, ostensibly for the benefit of your audience.

Call out partial answers as soon as you can: people will remember that you began to speak almost immediately, and will forget that you had not yet finished calculating. If challenged with $\sqrt{3.14}$, you can immediately say «One point seven ...», then add another digit soon enough. The conclusion, with three digits

1 From a song in the musical *Chicago*.

delivered at the same time on a triumphant tone, will make up for the time spent on calculation. Similarly, when calculating a common logarithm, you can immediately give the characteristic: «What is log 314, you ask? Two point ...». When you give a partial answer, let your intonation show that you have not finished.

Recite answers in full numeric form, which is longer, more formal, and more impressive: say not «thirteen eighty-four» but «one thousand three hundred eighty-four».

If you calculate days of the week, you can gain speed by learning a code that will stand in for the present year; then you can compute a day of the week by adding the code for the month and the day, subtracting 1 if the date falls in the first two months of a leap year, and taking the remainder from division by 7. After all, when in the presence of others, you are most likely to need a date in the present year. For one in the following year, add 1 to your code (2 for a leap year).

I am writing these words in 2024, for which year the code is the remainder from $(6 + 24 + 6)/7$, namely 1. By memorising that code, I can give the day of the week for any day in 2024 before anyone can reach for a calendar. How about 29 February? To the code 1, add 3 for the month and 29 for the day; subtract 1 because the date falls within the first two months of a leap year. Then $32/7$ gives a remainder of 4, so the day is a Thursday. And 9 October? We get $(1+0+9)/7$, whose remainder is 3: Wednesday. Is someone already asking about dates in 2025? Add 1 to the code for 2024, getting 2, and proceed with the calculation.

Performing in front of an audience, you might use a large calculator, perhaps on an overhead projector, to show the correctness of your calculations. Instead, you might assign a few members of the audience to check your work on calculators (such as the one found on almost every mobile telephone). Even if your calculation is wrong, those with the calculators may fail to correct you, on the assumption that they themselves made a mistake.[82]

If you are caught in an error, admit it but minimise it, perhaps by quoting Virgil: «Quandoque bonus dormitat Homerus» ‘Even the great Homer sometimes nods off’. Moving on, you might ask for a «harder» problem—one that may not be harder after all.

I make mistakes, and so will you—especially under the pressure of a public performance. Prepare a graceful way to recover, one that will find favour with the audience.

In everyday life, you will have many opportunities to perform a computation for which others would have sought a calculator. I am inclined to answer any expression of amazement by downplaying the feat: «Oh, that’s nothing. Shall I calculate a square root for you? a logarithm?» You may prefer to take a bow, or even to affect mannerisms that make the feat look more difficult than it is.

Just before this book went to press, opposing counsel wondered aloud in court how much \$128 per month for 28 months was. Within a second or two, I gave the answer: \$3 584. She retrieved her calculator and confirmed my answer. «Very impressive, Mr Horne», said the judge. It was easy: $128 \cdot 28 = 100 \cdot 28 + 28^2 = 2\,800 + 784 = 3\,584$. I did not, however, reveal

2 Arthur Benjamin’s mistakes during performances available on YouTube usually pass without comment, presumably because the people with the calculators doubt themselves or expect that someone else will object.

how easy the calculation was. To everyone's amusement, I mentioned that this book would be coming out in a few weeks and that anyone interested in learning my methods was welcome to buy a copy.[83]

83 See Appendix E for a joke about lawyers' notoriously poor understanding of mathematics.

Chapter 24

Next steps

Xúnzǐ, a Chinese philosopher who flourished in the third century BCE, made a figurative observation that is still a well-known proverb today: 青出于蓝而胜于蓝 'Blue dye comes from the indigo plant but is bluer than the indigo plant'. The meaning is that the student may come to surpass the teacher. I fervently hope that you will surpass me and carry mental arithmetic to new heights.

Having reached the end of this book, you may wonder what to do next. This chapter offers a few ideas for your consideration.

Everyday use

Put your hard-acquired skills to use! You can practise them very effectively by using them in real life, rather than resorting to calculators or other aids. Why not? Use what you have learnt.

I seldom use a calculator. My HP-15C from the 1980s has not had batteries in more than a decade. Most of the time I can do what I need in my head, even if it involves logarithms, exponentials, roots, or trigonometric functions. There is satisfaction in getting an exact answer, or a good approximation, unassisted.

Look for opportunities to practise when you would otherwise waste your time. I often square the numbers of licence plates on passing cars while driving, or calculate their logarithms. Why not extract a square root or two while waiting at the bank or the physician's office?

Refined methods

You may refine the methods presented in this book; you may even develop superior methods of your own. To do so, you will probably require a good knowledge of mathematics—algebra at a minimum, and trigonometry or the calculus for some of the fancier functions. You may advance the state of knowledge in this area. If so, please tell me: I am always eager to learn of better ways to perform mental arithmetic.

Some of the works cited in the bibliography present methods of mental calculation that do not appear in this book. You may wish to explore those methods for possible adoption into your repertoire. They may give you ideas for developing better methods of your own.

Competitions

If you develop enough speed, you may join a regional or international competition. Different categories exist: adding numerous digits, multiplying large numbers, extracting roots, performing calendrical calculations, and others.

I have never taken part in one of these competitions, nor do I intend to; I am content with using mental arithmetic for my own purposes and sharing its possibilities with people like you. If now and then someone admires my facility, well and good. I do not, however, clamour for attention. Many people do, with the result that there is a great deal of hucksterism in the field of mental arithmetic: people claim divine powers or pass memorised numbers off as spontaneous calculations. Celebrity seems to attract self-aggrandisers as dung attracts flies.

Although I do not seek recognition or claim to be an expert at mental arithmetic, those of a more competitive or ambitious bent are welcome to try their skills against those of others.

Broader study of mathematics

I do hope that your study of mental arithmetic will inspire you to learn more about mathematics. With its explanations of the reasons behind its techniques, this book gives you a good start.

Mathematics is often supposed to be the exclusive demesne of a skilled élite, a field too abstruse and difficult for the general public. Women in particular are still discouraged from mathematical pursuits. Anyone, however, can learn math through study and diligent application. It is at bottom nothing but reasoning, dealing as it does with pure concepts that often turn out to have important practical applications. One does have to approach it systematically, mastering basic skills before progressing to anything that depends on them.

An entire course in conventional arithmetic, abbreviated for intelligent adults, is found in the first chapter of this book. If you learn nothing else from this book, please solidify your knowledge of ordinary arithmetic. You will then have a good foundation for the study of algebra and other subjects, if you are interested in them. You will also be more capable of functioning in a world full of data. Perhaps you will experience something beautiful.

Appendix A

Bibliography

VIRTUOSI-TO-BE may profit from the works listed in this annotated bibliography. Unfortunately, much of the literature on mental arithmetic is of poor quality, consisting of little but elementary notions and gimmickry; some of it even degenerates into chauvinism and fraud.[84] Very few authors go far beyond addition, subtraction, multiplication, and division. The reader seeking further study in the field of mental arithmetic or the closely allied field of memory would do well to consider the works described below.

Aitken, AC. «The Art of Mental Calculation: With demonstrations» (1954) The Transactions of the Society of Engineers (Incorporated) 295.

A native of New Zealand, Alexander Aitken was not only an accomplished mathematician but also perhaps the greatest mental calculator of all time. This article, effectively the minutes of

4 I am thinking in particular of «Vedic mathematics», a false concept that has attracted attention because of its associations with Indian mysticism. India has undeniably played a major rôle in the history of mathematics, notably by inventing our «Arabic» numbers, properly called Hindu-Arabic. The notion that the Vedas contain an entire school of arithmetic, however, is simply false and should not be repeated.

a meeting of engineers at which Aitken spoke, offers rare insight into Aitken's methods. Squaring, at least of three- and four-digit numbers, he performed with the general technique taught in this book.

Especially worthy of reading, but in places insufficiently developed, is Aitken's manner of calculating square roots, which exploits averaging with astounding efficacy. Ron Doerfler has done the world a service by explaining Aitken's technique.[85]

Aitken «deplore[d] all tendency to surround the faculty [of mental calculation] with an aura of the marvelous», as do I: mental calculation is open to anyone who can learn proper technique and apply herself diligently. Aitken freely admits that he was undistinguished at arithmetic as a child and that he began to pursue mental arithmetic only when he was nearly fourteen years old, after an algebra teacher mentioned in passing that the difference of squares, no mere statement about numbers in the abstract, had practical application for mental squaring of specific numbers. He developed most of his ability during the following four years. He shared my opinion—the central tenet of this book—that mental arithmetic, like any other skill, can be learnt and developed.

This article is hard to find; you may need to consult an academic library. I found an offprint, with handwritten annotations presumably by Aitken himself, in the collection of Aitken's

85 See *Dead Reckoning* at 80–85. The chapter on estimation in the present book introduces the fact that, if x is a decent estimate of $\sqrt{R}$, a better estimate is the average of x and R/x, namely $(x + R/x)/2$. Typically the average can be obtained much more easily, by adding half of the difference of x and R/x to the lower of the two. The process can be repeated, with the new estimate standing in for x, to obtain better and better approximations of the root. This use of the average is equivalent to the Newton–Raphson method of calculating square roots, which I introduce for obtaining six digits with relatively little effort. If you are curious about extending the method to many more digits, please read Doerfler's account of Aitken's approach.

papers that is housed at the library of the University of Edinburgh, where he taught.

❧

Benjamin, Arthur & Michael Shermer. *Secrets of Mental Math: The mathemagician's guide to lightning calculation and amazing math tricks* (New York: Three Rivers Press, 2006).

Written by a professor of mathematics who exhibits mental arithmetic in the style of a magic act, this may be the most popular work in the field of mental arithmetic. Most of the methods are quite conventional. Benjamin discusses the practical problem of remembering numerous digits; he may be one of the first authors to explain the use of the mnemonic major system as an aid to mental arithmetic, although I have been using it for that purpose since learning about it in the early 1980s. Multiplication and squaring are extensively covered; division is insufficiently developed. A simple approximation of square roots is presented, as is a method for calculating the day of the week; but logarithmic, exponential, and circular functions are not introduced. Some flashy recreational mathematics, including a formula for generating a type of magic square with special properties, rounds out the book. Meant for the general reader.

❧

Doerfler, Ronald W. «A Method for 4×4 Digit Mental Multiplications» (2006), online: ⟨https://deadreckonings.com/mental-calculation-papers/⟩.

The method of this paper is summarised in the chapter on advanced multiplication. The paper explains why the method works and how it was derived.

❧

Doerfler, Ronald W. *Dead Reckoning: Calculating without instruments* (Houston: Gulf Publishing, 1993).

My favourite contribution to the literature on calculation, this important book offers mathematically sophisticated readers the fruits of Doerfler's research and of some methods that he himself developed or improved. He correctly states (at 5) that no comparable book exists: no other author has so carefully combed the literature for numerical methods adapted to practical manual or mental work.

The book passes quickly over addition, subtraction, multiplication, and division to devote space to factorisation (apparently one of the author's special interests), roots, logarithms, exponentials, and trigonometric and inverse trigonometric functions—topics seldom treated seriously, if at all, in works on mental arithmetic. Most of Doerfler's methods for these fancier calculations differ from mine; the reader with a solid grasp of algebra and at least basic knowledge of the differential calculus may gain from exploring his. Of especial interest is his iterative algorithm for extracting square roots to any desired precision.

Numerous errata, published in a list on the Internet,[86] unfortunately mar the text; I have corrected my copy by hand. It is to be hoped that Doerfler will issue a new edition of this important work, fully corrected but also expanded with his more recent discoveries.

Doerfler, Ronald W. «Developing a "Number Sense" through Mental Multiplication» (6 November 2006), online: ⟨https://deadreckonings.com/mental-calculation-papers/⟩.

This paper convincingly makes the case that mental multiplication, beyond being beneficial in its own right, fosters an under-

86 Online: ⟨https://deadreckonings.com/mental-calculation-papers/⟩.

standing of numbers that the mechanical algorithm of manual multiplication, a banausic manipulation of digits, arguably impedes. Most of the techniques described also appear in this book, in the chapters on mutliplication or squaring: they are very old, for the most part, and all stem from simple algebra. A few techniques that work best for small classes of numbers are included not as practical means of mental multiplication but as illustrations of the properties of numbers and the curious discoveries that await the inquisitive student of mathematics.

Doerfler, Ron. «Fast Approximation of the Tangent, Hyperbolic Tangent, Exponential and Logarithmic Functions» (27 June 2007), online: ⟨https://deadreckonings.com/mental-calculation-papers/⟩.

Published after *Dead Reckoning*, this paper introduces four-digit approximations of the four named functions that are designed for quick mental work. Although few readers will want a mental algorithm for the hyperbolic tangent, the one given here lends itself to the calculation of the natural exponential function and the natural logarithm as well. Highly recommended to those who frequently compute e^x or $\ln x$, and certainly to those who need a good method for $\tanh x$.

Doerfler, Ron. «The McIntosh–Doerfler Algorithm for Exponentials» (17 June 2021), online: ⟨https://deadreckonings.com/mental-calculation-papers/⟩.

John McIntosh and Ron Doerfler developed a clever way to calculate 10^x to five digits' accuracy with only two memorised numbers: $\log 2$ and $\log 3$. The values of those two logarithms happen to make this method practical. Examples illustrate the use of the method. See also John McIntosh's article «Exponentials», online: ⟨https://www.urticator.net/essay/6/641.html⟩.

Hunter, Ian ML. «An Exceptional Talent for Calculative Thinking» (1962) 53 : 3 British Journal of Psychology 243.

Hunter, who happened to teach at the same institution as Aitken (the University of Edinburgh), interviewed him and analysed his manner of calculation from a psychological perspective, with a focus on conversion of fractions to decimal form.

Despite being generally underwhelmed with psychobabble, I find that this article gives a fair analysis—much fairer than the reference to «talent» in the title suggests.[87] Hunter observes that Aitken perceived useful «attributes» of numbers, such as their prime factors, and was able to select a convenient manner of proceeding where most people would have only one—long division. Moreover, Aitken's strong memory for numbers enabled him to store and recall intermediate results that most people would quickly forget. (Often he did not know what he was calculating and what he was simply drawing from memory.) As a teenager, he spent four years on diligent practice that he described as «mental Yoga» (at 252).

87 See Geoff Colvin, *Talent Is Overrated: What* really *separates world-class performers from everybody else* (New York: Portfolio, 2008). «Talent», as innate ability, hardly exists; what underlies excellence in any field—academic, athletic, artistic, professional—is deliberate practice, a protracted programme of identifying and correcting weaknesses. I go further: «talent» is an excuse that people use to justify their own inferior performance. *Of course she is an expert: she has talent. I don't, so I cannot hope to achieve her heights.* Nonsense! I have not yet seen anyone who could perform Chopin's études after a month at the piano or qualified for the Bolshoi weeks after first donning a pair of dance slippers. Consistently one finds that those who excel—including Aitken—put thousands of hours into deliberate practice.

Particularly curious was Aitken's method of calculating $\frac{1}{697}$ (at 250). At first he gave up after 0.001 4, which was as far as he could carry long division. (He could easily have done better by calculating $\frac{1}{700}$ + 0.004%.) After stating that he might have practised had he expected a «very difficult» three-digit divisor, he noticed that $697 = 17 \cdot 41$, and he proceeded by calculating $\frac{1}{41}$ and dividing by 17 as he went, obtaining 0.001 434 72 before giving up. Later, he wrote to tell Hunter that he was «really disappointed» in himself: he felt that he should have been able to keep 02439, the repeating period of $\frac{1}{41}$, in mind and divide by 17 for many more decimal places. (The repeating period of $\frac{1}{697}$, as he pointed out, is 80 digits long.)

I would simply have divided 1 by 700 while adding 3 times the new digit of the quotient to each remainder: 0.001 remainder $300 + 3 \cdot 1 = 303$, with a 0 brought down makes 3030; then 4 remainder $230 + 3 \cdot 4 = 242 \longrightarrow 2420$; then 3 remainder $320 + 3 \cdot 3 = 329 \longrightarrow 3290$.... Even bringing down the zeroes is not necessary if one divides by 70 instead (which amounts to dividing by 7 but avoiding error when the remainder is below 70 and the quotient is therefore 0). I am surprised that Aitken resorted to long division and could not continue it for more than two digits.

Lorayne, Harry & Jerry Lucas. *The Memory Book* (New York: Basic Books, 1964).

A classic twentieth-century work on memory, this book presents among other things the mnemonic major system (without calling it by that name) and shows applications of it to the remembering of facts that on their face appear to have little to do with numbers. Lorayne too was a showman: he specialised in remembering the names of everyone in an audience that might include hundreds of people. Directed to the general reader who is eager to remember things more effectively in daily life.

Menninger, Karl. *Calculator's Cunning* (New York: Barnes & Noble, 1974).

Originally published in German in 1931 (as *Rechenkniffe: Lustiges und vorteilhaftes Rechnen*), this book is still in print. It teaches sensible methods for addition, subtraction, multiplication, division, checking of calculations, and estimation, while explaining why those methods work. Meant for the general reader.

Smith, Stephen B. *The Great Mental Calculators: The psychology, methods, and lives of calculating prodigies past and present* (New York: Columbia Univ Pr, 1983).

Most people too readily suppose that skill at mental arithmetic is a divine gift, something beyond themselves. Mental calculators whose reputations are enhanced by this confusion are often happy to perpetuate the deception.

Smith's sober account of famous mental calculators of the past few centuries differs refreshingly from the breathless and credulous accolades so frequently encountered. Far from being endowed with superhuman or preternatural abilities, these calculators simply deploy methods optimised for mental arithmetic, such as those taught in the book that you are now reading. Like people who excel in any other field, mental calculators tend to be monomaniacs, dedicating uncommonly much time and practice to their endeavour. Moreover, Smith candidly observes that the accounts of their successes are often vague, uncorroborated, or exaggerated.

Smith also presents some of the methods used by mental calculators. By now you will know that they usually differ greatly from those of conventional arithmetic.

Long out of print, this book is available from many libraries. It is the best—nearly the only—history of mental calculators, offering insight into the reality of what they achieved and how they achieved it.

Sticker, Henry. *How to Calculate Quickly* (New York: Dover Publications, 1955).

Still in print, this book features drills that over time cultivate speed and expand the capacity for carrying numbers in the head. The reader who works through the drills in order, even for just a few minutes a day, will progress at mental arithmetic.

Appendix B

The binomial theorem

Statement

A *binomial* is nothing but the sum of two terms, $x + y$. Raising a binomial to a power might seem to entail a lot of tedious algebraic manipulation. The important *binomial theorem*, however, enables us to calculate $(x + y)^n$ for any non-negative integer n in a straightforward way:[88]

$$(x + y)^n = \sum_{k=0}^{n} \binom{n}{k} x^{n-k} y^k.$$

Before we prove and apply this theorem, an explanation of the new notation above is in order.

Definitions and notation

The *factorial* of a positive integer k, written $k!$, is the product of all integers from k down to 1. Thus $4! = 4 \cdot 3 \cdot 2 \cdot 1 = 24$. By definition, $0! = 1$.

---

88 The binomial theorem has been extended: the great English mathematician and natural scientist Isaac Newton generalised it to real powers, and it has been further generalised to complex powers. The generalised forms use an infinite series in place of a finite sum. Discussion would take us too far from our purpose; if you are interested, consult any text on the binomial series. In this appendix, only non-negative integral powers are considered.

The *binomial coëfficient*, written $\binom{n}{k}$ (or in an older style ${}_nC_k$) and called «n choose k», is defined as

$$\binom{n}{k} = \frac{n!}{k!\,(n-k)!}.$$

Despite the presence of a fraction, the binomial coëfficient is always an integer: the factors in the denominator all cancel out.

Summation is just a simplified notation for adding multiple terms without writing them all out. The notation $\sum_{k=0}^{n}$ means that the term following it is used with each integer k from 0 to n inclusive and that all such terms are added up. For example,

$$\sum_{k=3}^{5} \frac{k}{k+2} = \frac{3}{3+2} + \frac{4}{4+2} + \frac{5}{5+2}.$$

Thus

$$\sum_{k=0}^{n} \binom{n}{k} x^{n-k} y^{k} = \binom{n}{0} x^{n-0} y^{0} + \binom{n}{1} x^{n-1} y^{1} + \binom{n}{2} x^{n-2} y^{2} + \cdots$$
$$+ \binom{n}{n-1} x^{n-(n-1)} y^{n-1} + \binom{n}{n} x^{n-n} y^{n}.$$

Why the binomial theorem is true

The following combinatoric argument lacks the rigour of a mathematical proof but affords an intuitive understanding of the binomial theorem and its correctness.

The binomial coëfficient $\binom{n}{k}$ gives the number of combinations of k distinct items that can be selected from a set of n items. For example, the number of hands of 5 cards that can be dealt from a deck of 52 is

$$\binom{52}{5} = \frac{52!}{5!\,47!}$$

$$= \frac{52 \cdot 51 \cdot 50 \cdot 49 \cdot 48 \cdot \cancel{47!}}{5 \cdot 4 \cdot 3 \cdot 2 \cdot \cancel{47!}}$$
$$= 52 \cdot 51 \cdot 10 \cdot 49 \cdot 2$$
$$= 2\,598\,960.$$

The reason is that there are 52 choices for the first card, 51 for the second (all but the card just chosen as the first card), and so on, down to 48 for the fifth card; thus there are $52 \cdot 51 \cdot 50 \cdot 49 \cdot 48 = 311\,875\,200$ ways to deal five cards. That figure, $52!/47!$, overstates the number of *hands*, however, because a hand is the same irrespective of the order in which its cards were dealt. We must therefore divide by the number of orders. Since there are 5 positions for the first card, 4 remaining positions for the second card, and so on, the number of possible orders is 5!. Therefore, the number of hands is $52!/(5!\,47!) = \binom{52}{5}$.

Expanding $(x+y)^n$ yields a sum of terms each of which consists of n factors that are either x or y. For example,

$$\begin{aligned}
(x+y)^3 &= (x+y)(x+y)(x+y) \\
&= (xx + xy + yx + yy)(x+y) \\
&= xxx + xxy + xyx + xyy + yxx + yxy + yyx + yyy \\
&= xxx + (xxy + xyx + yxx) + (xyy + yxy + yyx) + yyy \\
&= xxx + 3(xxy) + 3(xyy) + yyy \\
&= x^3 + 3x^2y + 3xy^2 + y^3.
\end{aligned}$$

In each term, there are k factors of y and $n-k$ factors of x, for a total of $n-k+k=n$ factors in all. Thanks to the commutative property of multiplication, terms with the factors in a different order are equivalent: for instance, $xxy = xyx = yxx$. How many such combinations are there? If y is raised to the power k, there are $\binom{n}{k}$ combinations of factors that make $x^{n-k}y^k$, since there are $\binom{n}{k}$ ways to choose k factors of y from a set of n and only one way to make up the remaining factors (namely, each of them is x). Therefore, the sum of the combinations that equal $x^{n-k}y^k$

is $\binom{n}{k}x^{n-k}y^k$, and the sum of all terms containing a power of y is $\sum_{k=0}^{n}\binom{n}{k}x^{n-k}y^k = (x+y)^n$.

Pascal's triangle

The binomial coëfficients form a pattern called *Pascal's triangle*, named after seventeenth-century French mathematician Blaise Pascal. Each coëfficient is the sum of the two above it, diagonally to the left and right.

$n = 0$							1						
$n = 1$						1		1					
$n = 2$					1		2		1				
$n = 3$				1		3		3		1			
$n = 4$			1		4		6		4		1		
$n = 5$		1		5		10		10		5		1	
$n = 6$	1		6		15		20		15		6		1

Each row of the triangle contains all of the binomial coëfficients for its value of n. For example, the row for $n = 4$ has 1 4 6 4 1; therefore, $(x+y)^4 = x^4 + 4x^3y + 6x^2y^2 + 4xy^3 + y^4$.

Observations

As one would expect, the expanded binomial is symmetrical: the same coëfficients occur on each side. Indeed,

$$\binom{n}{n-r} = \frac{n!}{(n-r)!\,(n-(n-r))!} = \frac{n!}{(n-r)!\,r!} = \frac{n!}{r!\,(n-r)!}$$

$$= \binom{n}{r}.$$

In particular,

$$\binom{n}{0} = \binom{n}{n} = \frac{\cancel{n}!}{\cancel{n}!\,0!} = 1$$

and

$$\binom{n}{1} = \binom{n}{n-1} = \frac{n!}{(n-1)!\,1!} = \frac{n \cdot \cancel{(n-1)!}}{\cancel{(n-1)!}\,1!} = n.$$

Therefore, the coëfficient of the terms x^n and y^n is 1, and the coëfficient of the terms $x^{n-1}y$ and xy^{n-1} is n.

❧

Consider

$$(1+x)^n = 1 + nx + \frac{n(n-1)}{2}x^2 + \cdots .$$

If $|x| < 1$, the higher powers of x are smaller ﬅill. Indeed, if $|x|$ is sufficiently small, we can ignore the powers from x^2 up and use the approximation

$$(1+x)^n \approx 1 + nx.$$

For example, $1.01^3 = (1+0.01)^3 \approx 1.03$.

Appendix C

Taylor series

This appendix requires a knowledge of the differential calculus. If you lack such knowledge, you may still profit from the discussion of Euler's Theorem at the end.

Let f be an infinitely differentiable function, with a positive radius of convergence, defined by the following power series:[89]

$$f(x) = \sum_{n=0}^{\infty} c_n x^n = c_0 + c_1 x + c_2 x^2 + \cdots .$$

Consider its first few derivatives:

$$\begin{aligned}
f'(x) &= c_1 + 2c_2 x + 3c_3 x^2 + 4c_4 x^3 \\
&\quad + \cdots + nc_n x^{n-1} + \cdots \\
f''(x) &= 2c_2 + 2 \cdot 3c_3 x + 3 \cdot 4c_4 x^2 \\
&\quad + \cdots + (n-1)nc_n x^{n-2} + \cdots \\
f'''(x) &= 2 \cdot 3c_3 + 2 \cdot 3 \cdot 4c_4 x \\
&\quad + \cdots + (n-2)(n-1)nc_n x^{n-3} + \cdots \\
f^{\text{iv}}(x) &= 2 \cdot 3 \cdot 4c_4 + 2 \cdot 3 \cdot 4 \cdot 5c_5 x \\
&\quad + \cdots + (n-3)(n-2)(n-1)nc_n x^{n-4} + \cdots
\end{aligned}$$

89 The notation $\sum$ is explained in the appendix on the binomial theorem.

Only the leading term of each derivative lacks a factor of x; therefore, we have

$$\begin{aligned} f(0) &= c_0 \\ f'(0) &= c_1 \\ f''(0) &= 2c_2 \\ f'''(0) &= 2 \cdot 3c_3 \\ f^{\text{iv}}(0) &= 2 \cdot 3 \cdot 4c_4 \\ &\vdots \\ f^{(n)}(0) &= n!\, c_n \end{aligned}$$

Hence

$$c_n = \frac{f^{(n)}(0)}{n!}$$

Therefore,

$$\begin{aligned} f(x) &= \sum_{n=0}^{\infty} \frac{f^{(n)}(0)}{n!} x^n \\ &= f(0) + f'(0)x + \frac{f''(0)}{2!}x^2 + \frac{f'''(0)}{3!}x^3 + \cdots . \end{aligned}$$

This expression gives the *Taylor series* for f.[90]

Taylor series often provide useful rational approximations of functions that are otherwise difficult to compute. Some Taylor series do not converge, or converge only within a small interval;

90 More correctly, it gives the *Maclaurin series*. The Taylor series at a is the same expression, but with x changed to $x - a$ and $f^{(n)}(0)$ changed to $f^{(n)}(a)$. The Maclaurin series is simply the Taylor series at 0. In practice, the Maclaurin series is often called the Taylor series.

some converge impractically slowly. Fortunately, the Taylor series for most of the elementary functions in this book do converge well and are therefore useful for manual or even mental arithmetic:

$$e^x = \sum_{k=0}^{\infty} \frac{x^k}{k!} = 1 + x + \frac{x^2}{2} + \frac{x^3}{6} + \frac{x^4}{24} + \cdots$$

$$\sin x = \sum_{k=0}^{\infty} \frac{(-1)^k x^{2k+1}}{(2k+1)!} = x - \frac{x^3}{6} + \frac{x^5}{120} - \frac{x^7}{5\,040} + \cdots$$

$$\cos x = \sum_{k=0}^{\infty} \frac{(-1)^k x^{2k}}{(2k)!} = 1 - \frac{x^2}{2} + \frac{x^4}{24} - \frac{x^6}{720} + \cdots$$

$$\tan x = x + \frac{x^3}{3} + \frac{2x^5}{15} + \frac{17x^7}{315} + \cdots$$

$$\arcsin x = x + \frac{x^3}{6} + \frac{3x^5}{40} + \frac{5x^7}{112} + \cdots$$

$$\arccos x = \frac{\pi}{2} - \arcsin x = \frac{\pi}{2} - x - \frac{x^3}{6} - \frac{3x^5}{40} - \frac{5x^7}{112} - \cdots$$

$$\arctan x = \sum_{k=0}^{\infty} \frac{(-1)^k x^{2k+1}}{2k+1} \qquad \text{for } -1 < x < 1$$

$$= x - \frac{x^3}{3} + \frac{x^5}{5} - \frac{x^7}{7} + \cdots$$

For many values, the Taylor series for the arctangent converges too slowly for practical use.

The natural logarithm has no Taylor series at 0, since its first derivative, $\frac{1}{x}$, is not defined at 0. Instead, this series centred at 1 converges over the interval $-1 < x \leq 1$:

$$\ln(1 + x) = x - \frac{x^2}{2} + \frac{x^3}{3} - \frac{x^4}{4} + \cdots.$$

Another good series for the natural logarithm, valid over the interval $-1 < x < 1$, is

$$\ln\left(\frac{1+x}{1-x}\right) = 2x + \frac{2}{3}x^3 + \frac{2}{5}x^5 + \frac{2}{7}x^7 + \cdots.$$

To compute $\ln y$ with this series, let $x = \frac{y-1}{y+1}$.

Taylor series with good convergence can often be truncated at just a few terms, at least for some values. Thus, when x is close to 0, we know that $\ln(1+x)$, $\sin x$, $\tan x$, $\arcsin x$, and $\arctan x$ can all be approximated as x; $\cos x$, as 1; $\arccos x$, as $\frac{\pi}{2}$; and e^x, as $1+x$. These conclusions follow from an examination of the Taylor series: the lower-order terms quickly become much smaller than x and can all be ignored.

For larger values, sometimes two or three terms of the Taylor series suffice. Often each subsequent term is easily computed from the previous one. For e^x, the first term is 1; the second is the first term times x divided by 1; the third is the second times x divided by 2; the fourth is the third times x divided by 3; and so on. Functions such as $\sin x$ and $\ln(1+x)$ can be computed similarly.

Although I promised in the first chapter not to introduce calculations with complex numbers, I am making an exception in this appendix to show you something that is well worth seeing.

The eminent eighteenth-century Swiss mathematician Leonhard Euler studied complex numbers when they were still poorly understood. (He himself committed basic errors that led him to adopt the symbol i in place of $\sqrt{-1}$.) The Taylor series had been developed only three decades before Euler wrote his monumental work on mathematical analysis *Introductio in analysin infinitorum*,[91] in which he famously introduced the number e and the natural logarithm.[92] Only three pages after that momentous introduction, he wrote:

> Post Logarithmos & quantitates exponentiales considerari debent Arcus circulares eorumque Sinus & Cosinus, quia non solum aliud quantitatum transcendentium genus constituunt, sed etiam ex ipsis Logarithmis & exponentialibus, quando imaginariis quantitatibus involvuntur, proveniunt, id quod infra clarius patebit.
>
> Now that we have considered the logarithms and exponentials, we should consider the circular arcs and their sines and cosines, for they not only constitute transcendental quantities of another type but also follow from the logarithms and exponentials themselves when these occur with imaginary quantities, as shall become clearer below.[93]

Expanding $(\cos x \pm i \sin x)^n$ for positive integral values of n by the binomial theorem, Euler found the Taylor series for $\cos x$ and $\sin x$:[94]

$$\cos x = 1 - \frac{x^2}{2!} + \frac{x^4}{4!} - \frac{x^6}{6!} + \cdots$$
$$\sin x = x - \frac{x^3}{3!} + \frac{x^5}{5!} - \frac{x^7}{7!} + \cdots$$

91 (Lausanne: Marc-Michel Bousquet, 1748).
92 *Ibid*, vol 1 at 89–90.
93 *Ibid*, vol 1 at 93. My translation.
94 *Ibid*, vol 1 at 97–99.

Using an approach of insufficient rigour, he proved what has come to be known as Euler's Theorem:[95]

$$e^{ix} = \cos x + i \sin x.$$

Instead of following Euler's approach, we derive this result from the Taylor series for e^x and the cyclical nature of the powers of i ($i^1 = i; i^2 = -1; i^3 = -i; i^4 = +1$; then $i^5 = i^4 i = 1i = i$ again, and the sequence of four repeats itself):[96]

$$\begin{aligned}
e^x &= 1 + x + \frac{x^2}{2!} + \frac{x^3}{3!} + \frac{x^4}{4!} + \frac{x^5}{5!} + \frac{x^6}{6!} + \cdots \\
e^{ix} &= 1 + ix + \frac{(ix)^2}{2!} + \frac{(ix)^3}{3!} + \frac{(ix)^4}{4!} + \frac{(ix)^5}{5!} + \frac{(ix)^6}{6!} + \cdots \\
&= 1 + ix - \frac{x^2}{2!} - i\frac{x^3}{3!} + \frac{x^4}{4!} + i\frac{x^5}{5!} - \frac{x^6}{6!} - \cdots \\
&= \left(1 - \frac{x^2}{2!} + \frac{x^4}{4!} - \frac{x^6}{6!} + \cdots\right) + i\left(x - \frac{x^3}{3!} + \frac{x^5}{5!} - \cdots\right) \\
&= \cos x + i \sin x.
\end{aligned}$$

Generations of mathematicians have marvelled at this important result and even interpreted it philosophically. It reveals the relationship, promised by Euler in the text quoted above, between the trigonometric (or circular) functions and the natural exponential function (and by implication the natural logarithm), in which that special number i, defined in this book's introductory chapter, plays a leading rôle. It also shows that the «imaginary» numbers and the other complex numbers, far from being

95 *Ibid*, vol 1 at 103–05.

96 Lest I stand accused of the same lack of rigour with which I just reproached noble Euler, I point out that the derivation is justified because the three series are all absolutely convergent. The commutative property of addition does not extend to infinite sums, so we have to be careful when rearranging infinitely many terms.

mere fanciful playthings of effete mathematicians, have genuine significance.

From Euler's Theorem can be derived the following (where x is a real number, in radians):

$$\sin x = \frac{e^{ix} - e^{-ix}}{2i}$$
$$\cos x = \frac{e^{ix} + e^{-ix}}{2}$$
$$\tan x = i \cdot \frac{1 - e^{2ix}}{1 + e^{2ix}}$$

Thus the sine, cosine, or tangent of a real number can be expressed simply in terms of exponentials of complex numbers.

Let us now take x to be π in the equation $e^{ix} = \cos x + i \sin x$. We obtain $e^{i\pi} = \cos \pi + i \sin \pi = -1 + 0i = -1$.[97] By adding 1 to each side of the resulting equation $e^{i\pi} = -1$, we obtain *Euler's identity*, beautiful enough to deserve large type:

$$e^{i\pi} + 1 = 0.$$

Euler's identity has justly been called the most remarkable equation in mathematics, for it creates a powerful and astonishing truth from nine elementary building blocks: the five most important numbers, namely 0, 1, e, π, and i; the fundamental operations of addition, multiplication, and exponentiation; and the concept of equality. Where but in mathematics do such great beauty and wonder emerge from seven little symbols?

97 Remember that the trigonometric functions in mathematics ordinarily use radians, not degrees. Since π radians is 180°, we have $\cos \pi = \cos 180° = -1$ and $\sin \pi = \sin 180° = 0$.

Appendix D

Formula for amortisation

In the chapter on financial calculations, the monthly payment A on an amortised loan with initial principal P repayable over a term of n months with a monthly interest rate of r (expressed as a decimal, not as a percentage) is said to be given by the formula

$$A = P \cdot \frac{r}{1 - (1 + r)^{-n}}.$$

We now proceed to derive this formula.

First, we define the *geometric series* and give an expression for its sum. A geometric series is the sum of terms each of which is the previous one multiplied by a factor r. If the first term is a, the second is ar, the third is ar^2, and so on; thus the geometric series is $a + ar + ar^2 + \cdots + ar^n$ for some integer n.

Let S_n represent the sum of the geometric series to n terms, and let r be any number other than 1. (For $r = 1$, the geometric series is just $a + a + a + \cdots = na$.) We have

$$\begin{aligned}
S_n &= a + ar + ar^2 + \cdots + ar^{n-1} \\
rS_n &= \phantom{a + {}} ar + ar^2 + \cdots + ar^{n-1} + ar^n \\
S_n - rS_n &= a - ar^n \\
(1 - r)S_n &= a(1 - r^n)
\end{aligned}$$

$$S_n = a \cdot \frac{1 - r^n}{1 - r}$$

This formula, $S_n = a(1 - r^n)/(1 - r)$, gives us a simple way to sum a geometric series. Multiplying numerator and denominator by -1, we obtain the alternative formulation

$$S_n = a \cdot \frac{r^n - 1}{r - 1},$$

which we use in this appendix.[98]

For an amortised loan, the principal (the balance of the loan) in any month is the previous month's principal plus the interest on that amount minus the monthly payment. At the outset, the principal is P: no interest has accrued, nor has any payment been made. After one month, the principal is $P(1+r)-A$: multiplying by $1+r$ adds the interest, and the payment A reduces the balance owing.

After two months, the principal is $\big(P(1+r)-A\big)(1+r)-A = P(1+r)^2 - \big(1+(1+r)\big)A$. After three months, the principal is $\big[P(1+r)^2 - \big(1+(1+r)\big)A\big](1+r) - A = P(1+r)^3 - \big(1 + (1+r) + (1+r)^2\big)A$. In general, after k months, the principal is $P(1+r)^k - \big(1 + (1+r) + (1+r)^2 + \cdots + (1+r)^{k-1}\big)A$.

98 For the sake of completeness, I should mention that there is also an *arithmetic series*, in which each term is the previous one *plus* a constant k: thus we have $a+(a+k)+(a+2k)+\cdots+\big(a+(n-1)k\big)$. The sum of this series is the number of terms times the average of the first and last terms. As a child, the great German mathematician and physicist Carl Friedrich Gauß (Gauss) surprised the teacher by adding the integers from 1 to 100—an assignment intended to keep the whole class busy—within minutes: he noted that $1 + 100 = 101$, $2+99 = 101$, $3+98 = 101$, and so on; thus he had 50 sums that equalled 101, and the total was therefore $50 \cdot 101 = 5\,050$.

Note that the coëfficient of A is a geometric series. Its sum is

$$1+(1+r)+(1+r)^2+\cdots+(1+r)^{k-1} = \frac{(1+r)^k-1}{(1+r)-1} = \frac{(1+r)^k-1}{r}.$$

At the end of the nth month, the loan has been paid off; the principal, therefore, is 0. We can use this fact to calculate A:

$$P(1+r)^n - A\cdot\frac{(1+r)^n-1}{r} = 0$$
$$A\cdot\frac{(1+r)^n-1}{r} = P(1+r)^n$$
$$A = P\cdot\frac{r(1+r)^n}{(1+r)^n-1} = P\cdot\frac{r}{1-(1+r)^{-n}}.$$

This is the formula for the monthly payment on an amortised loan that appears in the chapter on financial calculations.

Appendix E

Just for fun

Lawyers are notoriously ignorant of mathematics. Many of them took up law because a failing grade in Algebra I kept them out of medical school.[99]

Four lawyers on vacation stopped at a hotel. When the clerk offered them a room for $28, they asked how much they would each pay if they split the cost evenly.

The clerk reached for a sheet of paper. — There are 4 of you, and the cost is $28, so we must divide 28 by 4. We can do that with long division. I'm in law school, so I know everything. Let me show you how it's done.

$$
\begin{array}{r}
25 \\
4\,\overline{)\,28} \\
8 \\
\hline
20 \\
20 \\
\hline
\end{array}
$$

— Now, 4 doesn't go into 2, but 4 does go into 8. To be exact, 8 divided by 4 is 2. Then 2 times 4, that's 8. We subtract 8 from 8, getting 0, and bring down the 2. Now, 4 goes into 20, 5 times; and 5 times 4 is 20. Subtract, and it comes out even. That'll be $25 each, please.

99 In the interest of making full disclosure, I admit that I am a lawyer myself, though I hope that by now the reader will credit me with some mathematical ability.

The lawyers cheerfully paid $25 each and headed off to their room. One packed the towels into a suitcase; another looked for the raunchy passages in the Gideons' Bible. A third—the brains of the group—started to think, for a change, and decided that they might have been ripped off. The fourth happened to know a lawyer in the town and suggested that they seek a consultation.

— Well, colleagues, Mathematics graciously gives us not one but two ways to check the calculation: addition and multiplication. Let's try multiplication first.

$$\begin{array}{r} 25 \\ 4 \\ \hline 20 \\ 8 \\ \hline 28 \end{array}$$

— Each of you paid $25, and there are 4 of you. If the charge is correct, 25 times 4 will be 28. Let's try it and see. Well, 4 times 5 is 20, and 4 times 2 is 8. Add: 0 and 8 are 8, and then the 2... Yes, that's 28, just as expected.

— But since you've paid for the full hour, I may as well show you how to do it with addition.

$$\begin{array}{r} 25 \\ 25 \\ 25 \\ 25 \\ \hline 28 \end{array}$$

— We write 25 down 4 times and see whether the total is 28.

He touched his pencil in turn to each digit going up the right-hand column, then down the left-hand column:

— 5, 10, 15, 20, 22, 24, 26, 28. No, you weren't cheated at all.

COVER

The painting reproduced on the cover is *Устный счёт: В народной школе С.А. Рачинского* (Mental Arithmetic: In the public school of S A Rachinsky), produced in 1895 by Russian painter Nikolay Bogdanov-Belsky. Rachinsky, the teacher depicted at the left, abandoned a position as professor of botany in order to found a school for the poor children of the village of Tatevo, Smolensk. His innovative pædagogy incorporated mental arithmetic into the curriculum, using a textbook that he himself had written.

Having discovered the young Bogdanov-Belsky's ability as a painter, Rachinsky took him into his own house, taught him at the school, and paid for specialised training in art. This painting, part of the collection of the State Tretyakov Gallery in Moscow, can be seen as a testament to the artist's teacher and benefactor.

In the painting, the children have been asked to solve mentally the problem on the blackboard:

$$\frac{10^2 + 11^2 + 12^2 + 13^2 + 14^2}{365}.$$

A couple of methods come to mind. Knowing the squares, I would calculate it directly: $(12^2 + 14^2) + (11^2 + 13^2) + 10^2 = 340 + 290 + 100 = 730$, and $730 \div 365 = 2$. Although I have rearranged the terms for convenient addition, adding them in the order given turns out to be better: $10^2 + 11^2 + 12^2 = 365$, and $13^2 + 14^2 = 365$; thus we have $(2 \cdot 365) \div 365 = 2$.

Perhaps the teacher wanted an algebraic approach, however. We have

$$\frac{(12-2)^2 + (12-1)^2 + 12^2 + (12+1)^2 + (12+2)^2}{365}.$$

Furthermore,

$$\begin{aligned}(a-b)^2 + (a+b)^2 &= a^2 - \cancel{2ab} + b^2 + a^2 + \cancel{2ab} + b^2 \\ &= 2a^2 + 2b^2.\end{aligned}$$

Therefore, $(12-2)^2 + (12+2)^2 = 2 \cdot 12^2 + 2 \cdot 2^2 = 2 \cdot 12^2 + 8$, and $(12-1)^2 + (12+1)^2 = 2 \cdot 12^2 + 2 \cdot 1^2 = 2 \cdot 12^2 + 2$. The problem thus becomes

$$\begin{aligned}\frac{5 \cdot 12^2 + 10}{365} &= \frac{5 \cdot 144 + 10}{365} \\ &= \frac{720 + 10}{365} \\ &= \frac{730}{365} \\ &= 2.\end{aligned}$$

COLOPHON

The text of this book is set in EB Garamond, a carefully drawn, well-stocked, freely distributed implementation by Georg Mayr-Duffner of Claude Garamond's elegant typeface from the end of the sixteenth century.

The headings are set in **Futura PT**, Vladimir Yefimov's rendition of the classically proportioned geometric design released by Paul Renner in 1927. Strikingly different in character and era, it forms on the page with Garamond a rich counterpoint held together by similar proportions, particularly the classical Roman proportions of the capitals.

The lettrines that gracefully open the chapters are set in Goudy Initialen, based on a design by Frederic Goudy circa 1908. Each of the twenty-six lettrines appears exactly once in the text.

The author typeset this book himself with the excellent TeX system of Donald Knuth.

Made in United States
Troutdale, OR
11/03/2026

56847558R00179